LIFE INSIDE THE DEAD MAN'S CURVE

The Chronicles of a Public-Safety Helicopter Pilot

Kevin McDonald

© 2015 L&J Publications, Inc.
All Rights Reserved.

Except for the use of brief quotations in reviews or marketing materials, no part of this book may be reproduced, stored in a retrieval system, or transmitted in any form or by any means without the express written permission of L&J Publications.

First published by Dog Ear Publishing
4011 Vincennes Road
Indianapolis, IN 46268
www.dogearpublishing.net

ISBN: 978-1-4575-4493-4
Library of Congress Control Number: 2016930867

This book is printed on acid free paper.
Printed in the United States of America

Contents

Foreword by Captain George Galdorisi, USN (Ret.) vii

Author's Note xi

How to Fly a Helicopter xiii

In Memoriam (for Kristin McLain) xvii

Prologue 1

1. There I Was, ... 5
2. Lee Harvey Oswald, Kosmos Spoetzl, and a German Guy Named Moses 25
3. A Chronic Case of Aerodynamically Induced Paranoia 51
4. Too Young for the Job (Part One) 61
5. Too Young for the Job (Part Two) 79
6. Anatomy of a STAR Flight Pilot 107
7. Flying without Fear 125
8. Flying in Exile 143
9. The Leper Colony Anthology (Volume One) 167
10. The Leper Colony Anthology (Volume Two) 209
11. The Leper Colony Anthology (Volume Three) 223
12. Ejection Seats, Fires, and Flameouts 245
13. The Longest Night 263
14. Two Weeks in September 303

    Epilogue 321

    The Letter 325

    Acknowledgments 327

Foreword

By Captain George Galdorisi, USN (Ret.)

This is a book about flying—not just any flying—but the kind of seat-of-the-pants flying that harkens back to the days when the airplane was a novelty, and barnstorming pilots entertained millions across the country. It is the kind of flying where the pilot is truly "one with the machine," and the best pilot isn't the one who must think a lot about what he's doing with the controls. Instead, the machine simply responds to his subconscious.

It's also a book about a specific type of aircraft—helicopters—unquestionably the most unique flying machines to evolve since the Wright Brothers rose precariously above the dunes at Kitty Hawk. Because it *is* a book about helicopters, it is also a book about helicopter *pilots*, who are a unique breed, categorically set apart from their fixed-wing brethren. How unique?

FOREWORD BY CAPTAIN GEORGE GALDORISI

Here is what the iconic newscaster, Harry Reasoner, had to say about helicopters and their pilots back in 1971:

> The thing is, helicopters are different from planes. An airplane by its nature wants to fly, and if not interfered with too strongly by unusual events or by a deliberately incompetent pilot, it will fly. A helicopter does not want to fly. It is maintained in the air by a variety of forces and controls working in opposition to each other, and if there is any disturbance in this delicate balance, the helicopter stops flying immediately and disastrously. There is no such thing as a gliding helicopter. This is why being a helicopter pilot is so different from being an airplane pilot and why, in general, airplane pilots are open, clear-eyed, buoyant extroverts, and helicopter pilots are brooders, introspective anticipators of trouble. They know that if something bad has not happened, it is about to.

Fair enough. But this book, *Life Inside the Dead Man's Curve*, is not exactly the lament of a disillusioned pessimist. It is a true story of hard-earned fulfillment, written by a helicopter pilot who falls closer to the fixed-wing end of Harry Reasoner's spectrum, a point on the scale that lies somewhere between the optimist and the realist. That said, this book is about what is arguably the most challenging kind of flying on the planet—helicopters engaged in lifesaving missions—and it documents both the triumphs and the heartbreaking tragedies that are inevitably part of the landscape when you're flying an EMS helicopter. EMS is short for "Emergency Medical Service," and most of us are at least vaguely familiar with EMS helicopters. They are the lifesaving aircraft that arrive at the scene of a horrific car crash, or show up when a person in a remote area suffers a heart attack, or are dispatched when a climber falls in an otherwise inaccessible area.

FOREWORD BY CAPTAIN GEORGE GALDORISI

In a way that is not entirely different from military aviators, EMS pilots are called upon daily to be the difference between life and death; and there have been many books written about military aviation, the line of work in which I was engaged for thirty years. For two of those years, in the late 1980s, I commanded the *Battle Cats* of HSL-43, a Navy helicopter squadron based at NAS North Island, in San Diego. A young Kevin McDonald, the author, was one of my pilots back then. Kevin excelled as a naval aviator and subsequently found his calling as an EMS pilot in Travis County, Texas. This story is not only about his journey—it's a never-before-revealed look into the confidential world of EMS flying.

The title of this book, *Life Inside the Dead Man's Curve*, reveals a great deal. Military flying, even in combat, is governed by long lists of rules and regulations. Safety is paramount, and because of this, the list of "don'ts" for military pilots is ponderously long. For helicopters specifically, there are combinations of altitudes and airspeeds that you must avoid as a military pilot. If a fledgling military helicopter pilot can't (or won't) avoid these dangerous combinations and routinely flies on the wrong side of the "dead man's curve," his career can be terminated abruptly and even violently. But EMS pilots, by the nature of their mission, spend a *huge* amount of time inside the dead man's curve. This means that to rescue an injured person, retrieve an accident victim, or pluck a stranded hiker from an impossible situation, they must purposely fly on the wrong side of the dead man's curve, at an altitude and airspeed that affords them no chance of recovery if an engine hiccups, a gust of wind hits them the wrong way, or any one of a dozen untoward events happens.

With *Life Inside the Dead Man's Curve*, Kevin McDonald takes us deep inside this world in a brisk narrative that is both uplifting and frightening. The reader will "fly along" on EMS missions and learn what it's like to live inside that dead man's curve. The stories will leave you breathless, and you will also deep-dive into the human psyche of the unique professionals who hear the EMS calling. Read this book, and the next time you see an accident or hear a siren, you'll get a knot in your stomach, knowing that the victim's best—and often only—chance is that someone "shaking the sticks" of an EMS helicopter is not far away.

Life Inside the Dead Man's Curve puts you inside the cockpit of an Emergency Medical Service helicopter on life-or-death missions. It is difficult to sum this book up in a few paragraphs, but in writing this Foreword, I was constantly reminded of a quote by World War II Admiral William "Bull" Halsey, who famously said, "There are no extraordinary men. . . . just extraordinary circumstances that *ordinary* men are forced to deal with."

Life Inside the Dead Man's Curve takes you on an extraordinary journey with the otherwise-ordinary professionals who perform remarkable feats when they "shake the sticks" of EMS helicopters. Strap in, hold on—and be prepared for an emotional, adrenaline-charged ride.

Author's Note

This book contains firsthand accounts of actual events from my life as a public-safety helicopter pilot. Using official records, notes, and personal recollections, I have accurately narrated the stories to the best of my ability. With few exceptions, the people in the book are real, and no names have been changed. In order to protect the anonymity of certain individuals who didn't have an opportunity to collaborate in this effort, they have not been identified by name. In deference to them, and out of respect for the families of those who are deceased, descriptions of nonessential facts surrounding some of the incidents described in this book have been altered in an effort to prevent any unintended, circumstantial identification of the people who were there.

HOW THE CONTROLS WORK

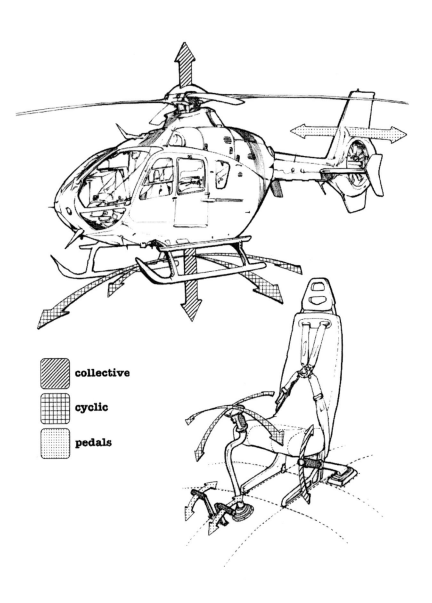

How to Fly a Helicopter

If you already know how to fly a helicopter, you can skip this section. For the rest of you, this is a condensed version of *Helicopters for Dummies*.

cyclic /s ī klik/

1. Located between the pilot's knees, it's equivalent to the stick in an airplane. With his elbow resting on his thigh, the pilot grips it with his right hand.
2. Used to roll the helicopter into turns during forward flight at altitude, it changes (independently and by varying amounts) the pitch on the main rotor blades as they *cycle* around the helicopter.
3. It can also be used to induce a nose-up or nose-down pitch.
4. Used to slide the helicopter left and right, forward and aft during hover.

collective /ku lek tiv/

1. Located immediately left of the pilot's seat, it's a horizontally mounted tube that articulates up and down. It *collectively* (equally) changes the pitch on each of the main rotor blades.
2. Gripping it with his left hand, the pilot raises it to add power and lowers it to reduce power, making it loosely equivalent to the throttle in an airplane.
3. Used to increase and decrease airspeed during forward flight at altitude.
4. During hover, raising and lowering the collective causes the helicopter to climb and descend respectively.

anti-torque pedals /anti-tork ped els/

1. Located on the cockpit floor, directly in front of the pilot, they increase and decrease pitch on the tail rotor.
2. The pilot uses them to rotate (yaw) the helicopter about the vertical axis. In other words, they're used to "point" the nose of the helicopter.

This book is dedicated to the former and current flight crews, mechanics, and administrative personnel of Travis County STAR Flight in Austin, Texas. You were the inspiration behind the narrative.

Kristin McLain

1969—2015

In Memoriam

During my last mission as a STAR Flight pilot, Flight Nurse Kristin McLain flew alongside me. On April 27, 2015, shortly before the completion of this book, Kristin lost her life during a hoist rescue, becoming the first Travis County STAR Flight crew member to die in the line of duty. A true professional in every sense of the word, her contributions to the STAR Flight program and the community she served will forever be remembered by those of us who worked beside her. Kristin was much more than a colleague—she was my friend.

Godspeed, Kristin

For Lee Ann and James

Maybe this will help to explain all the birthdays, Thanksgiving dinners, and Christmas mornings I missed while you were growing up.

Prologue

April 21, 2009

It was like trying to pick my way through a maze—only I was doing it in a helicopter, flying at 130 knots, barely 50 feet above the rugged Texas Hill Country landscape. The clouds were getting lower and lower, and every canyon into which we flew became a dead end.

We were desperately trying to reach the Lampasas River, where two men had driven their truck onto a flooded bridge and were now trapped in the rising water. The weather was deteriorating rapidly, and I had already considered aborting the mission several times during the flight from our home base in Austin. As I was hugging the deck, trying to find a way to reach our GPS coordinates and stay below the clouds, the surrounding hills were completely obscured.

Central Texas is notorious for flash floods, and over the years, my crew and I had rescued dozens of people from similar situations. I had been a STAR Flight pilot for two decades, and I had long ago learned that lousy weather was just part of the drill on rescue missions. Getting there was often the most difficult challenge.

We were definitely taking a circuitous route. Time after time, I saw an opportunity to turn toward our destination, only to be thwarted by the rising terrain in front of us. What looked like a clear path to the place where the men were trapped was repeatedly terminated in frustration at the end of a box canyon or a wall of dark clouds dipping all the way to the ground. It was looking like we would have to give up and turn around. Twenty minutes into the flight and forty miles from Austin, I wasn't sure we could even make it back home.

I was almost ready to call it quits and land when—out of nowhere—the river suddenly appeared below us. Although we were still more than twenty miles from the flooded bridge, I knew we might be able to follow the river and stay beneath the base of the black clouds.

Skimming only a few feet above the surface of the river, my biggest concern now was power lines. The visibility was so bad that it would be almost impossible to see them unless I spotted the towers from which they were suspended. Instead of concentrating my search straight ahead, I scanned back and forth to either side of the river, which was about as far as I could see in the poor weather conditions. Several times, my crew and I saw towers on one side of the river, but only once did the wires actually cross our path. I slowed our airspeed and cautiously flew between the towers. We never actually saw the power lines, but we knew they were directly over us, concealed by the low-hanging clouds.

Finally, we rounded a bend in the river and spotted the bridge where the two men were trapped in the middle of the rapidly rising river. There was a crew of emergency personnel at one end

PROLOGUE

of the bridge, but they had no way to reach the men, who had taken refuge in the bed of their pickup.

My crew chief quickly helped our rescuer rig himself to the hoist cable, and then, as I rolled onto my final approach to the truck, he moved the rescuer outside the aircraft. With the help of my crew chief's verbal commands, I maneuvered the helicopter into position while he lowered the rescuer to the flooded bridge. In only a few minutes, the first of the two men was hoisted from the back of the truck and into the helicopter. Shortly thereafter, we successfully hoisted the second man to safety as well—moments before the truck was swept from the bridge and carried downriver in the flood.

The men were grateful that my crew and I hadn't surrendered to the weather and aborted the flight. Neither of the two knew how to swim, and even if they'd been expert swimmers, their chances of survival would have been slim in the raging river.

1

There I Was, . . .

(If Only I Could Remember the Rest of the Story)

Almost every great barroom aviation story begins with these three words, "There I was." The expanded prelude normally includes something like:

> There I was, . . . upside down, in a flat spin, on fire, with three bogies on my tail.

Or, if the aviator telling the story happens to be a helicopter pilot:

> There I was, . . . tail rotor gone, cyclic jammed, almost out of altitude, airspeed, and ideas.

Well, it just so happens I *was* a helicopter pilot, and this is my story:

There I was, . . . at Brackenridge Hospital, in downtown Austin, Texas—sitting in front of the television. It was October 12, and the Texas Rangers were playing the Detroit Tigers in game four of the 2011 American League Championship Series. Five hours into the night shift, I was comfortably ensconced in a La-Z-Boy recliner inside the STAR Flight crew quarters.

With the score tied at three runs apiece in the top of the eleventh inning, Mike Napoli came to the plate with one out and runners at first and second. He lined an RBI single to centerfield, which gave the Rangers a four to three lead and brought up Nelson Cruz, who had thus far been the hottest hitter in the series. With two runners still aboard and still only one out, the count was one ball and two strikes to Cruz when he hammered a long fly ball to left-center field. I could tell it had a chance, so I came up out of the recliner yelling for the ball to "get out!"

As the ball cleared the wall and landed in the stands, I was pumping my fist and shouting, "Yes! . . . Yes! . . . Yes!" at the top of my lungs.

Cruz's three-run homer made it seven to three. Rangers pitcher Neftali Feliz sat the Tigers down in order in the bottom of the inning, giving Texas a three-games-to-one lead in the series. Three nights later, the Texas Rangers would win their second consecutive American League Pennant.

By that time, however, my career—and my life—had drastically altered course in a direction I could never have anticipated.

As the game ended, the on-duty flight nurse let me know she wasn't very happy with me. She had been snoozing in a back bedroom—right up until Nelson Cruz's three-run homer. As it turned out, it didn't really matter that my loud celebration had interrupted her sleep because—just as she was reading me the riot act—the pager sounded, which meant we were being tasked.

The pager was an amplified, audible alarm, which let us know we had been dispatched on a call (assigned to a mission). It began with rapid-fire, high-pitched chirping sounds, followed by a series of long, deafening tones, all of which combined to make it incredibly irritating to the human ear. I had really come to hate that sound over the past twenty years. It was especially excruciating if you happened to be sleeping when it went off, so in a way, I had actually done my nurse a favor by waking her up before we were dispatched.

Once the cacophony of tones had ended, the dispatcher informed us we were being assigned to a neonatal flight with a "specialty team" from St. David's, another nearby hospital. From our home base at Brackenridge, we would be flying to St. David's, where we would pick up a team of nurses from their NICU (neonatal intensive care unit), then fly them sixty miles east to the hospital in La Grange (the same town where "they got a lot of nice girls" in the ZZ Top song). Once we were in La Grange, the team would spend a couple of hours stabilizing the patient, a premature baby, at which time we would fly the team and the infant back to the NICU at St. David's in Austin.

On any other mission, we would be lifting off in the helicopter within five minutes from the time the pager went off, but specialty team flights were different. Before we could launch, there were always a few preflight administrative duties that had to be completed by the medical crew.

This gave me just enough time to scan the pantry. Hopefully, I'd find a snack, and I could make short work of it on my way up to the helipad. These missions usually took several hours to complete, and since it was already close to midnight, I knew there would be slim chow pickings at the hospital in La Grange. Unfortunately, the choices were equally lean in our pantry, so I grabbed a Diet Coke from the refrigerator and guzzled it down on my way out the door and up the stairs.

Because I was picking up an extra shift that night, I wasn't flying with my regular crew. My flight nurse, whom I had unwittingly annoyed just moments earlier, was relatively new. Kristin McLain was, at that time, the only female crew member at STAR Flight. "STAR" stood for Shock, Trauma, Air Rescue, and the physical demands placed on STAR Flight's medical crews were much more challenging than those required by most air ambulance programs. Not only were they tasked with providing prehospital medical care on EMS (emergency medical service) flights, the medics and nurses who flew with STAR Flight also functioned as rescue swimmers, hoist operators, and crew chiefs on a wide range of public-safety missions.

Kristin McLain had proven herself extremely competent in all of these roles. Just a month earlier, she and I had flown together on a rare nighttime fire-suppression mission. It had been an especially challenging operation, and Kristin—flying as my crew chief—had acquitted herself quite well, this despite working under some very difficult conditions. Even though she may have been slightly aggravated with me on this particular night, I still enjoyed flying with her.

My paramedic was Bill Hanson. Because we only took one STAR Flight crew member on specialty team flights, and it was Kristin's turn in the rotation, Bill would be sitting this one out at Brackenridge. Although he wasn't part of my regular crew, I was never disappointed when we were scheduled together. A U.S. Army veteran and a bit of a renaissance man, Bill was able to converse intelligently on a vast array of subjects. This helped pass the time on our long, twelve-hour shifts. Bill was from New Hampshire, and he liked to keep me entertained with his slightly left-of-center political views. Like most pilots, I'm somewhat conservative when it comes to ideology, and Bill was one of those rare individuals with whom you could disagree and still have a friendly, intelligent discussion about politics. That night, however, I wasn't so much interested in convincing Bill Hanson he was wrong about supply-side economics—I just wanted him to make a doughnut run so there'd be something to eat when I returned from La Grange.

While I was sitting in the cockpit, waiting for Kristin to finish making all the preflight arrangements with the NICU team, Bill came up to the pad and assumed his post next to the external power cart. When we were at Brackenridge, we always used external power during startups to conserve the helicopter's internal battery power.

"Is there anything you need?" Bill asked me.

Without hesitating, I answered his question. "That's affirmative, Bill. You can go get a dozen doughnuts and have them waiting for me when we get back."

"You know I can't do that," he said, laughing at my request. "If the boss calls, and I'm not here, I'll be in big trouble."

"It's after midnight, Bill. Nobody's going to call, and the doughnut shop is only ten minutes from here. You got nothin' else to do while we're gone. Man up!"

"Can't do it," he said again.

I hit the starter just as Kristin emerged from the stairwell. Then, making myself heard over the noise from the turbine spooling up, I yelled at Bill one last time.

"If you're a team player, Bill, there'll be doughnuts in the crew quarters when I get back!"

Bill just smiled, put his helmet on, and waited for me to give him the signal to disconnect the external power cart. As soon as the first engine was at idle, I gave him a thumbs-up. As I watched Bill flash a salute and roll the cart away, I was pretty sure there weren't going to be any doughnuts waiting for me when I got back from La Grange.

I started the second engine, and as Kristin finished strapping into the copilot seat, I rolled both throttles to the full-open position and finished my takeoff checks. During flights with no patient on board, it was standard procedure to have a medical crew member in the cockpit, and because the NICU team would be caring for the patient during the return flight from La Grange, Kristin would be up front with me for the entire mission.

"We good to go?" I asked, scanning the cockpit gauges and checking for caution lights.

"Good to go," she replied.

With that, I lowered my night-vision goggles and raised the collective until we were hovering just a few feet above the elevated, one-story pad. After one final cockpit check, I eased the cyclic forward and added just enough power to slide off the pad and start our climbout. Because it was such a short flight, we only had time to climb a few hundred feet before beginning our approach to the rooftop at St. David's.

As I completed my landing checks, I could see the team of two NICU nurses getting ready to roll their isolette up the long ramp to the pad. The isolette was an incubator, mounted to a gurney to make it portable. The gurney was equipped with collapsible legs to facilitate sliding it into our helicopter, where it could then be secured to the floor. The clear plastic dome, along with the rest of the equipment mounted to the gurney, provided a controlled, sterile environment for transporting a newborn infant. The entire assembly weighed about 350 pounds, so even though we were tasked with transporting a baby, we would actually be adding the weight of two adults to the back of the helicopter.

Once we were on short final, about a hundred yards from the elevated pad, I began slowing our approach speed and arresting our descent rate. I raised the collective and eased back on the cyclic, and then, just as we were about to come to a hover over our landing spot, I lowered the collective a little and let the helicopter settle onto the pad. As she unstrapped to get out and greet the team, Kristin reminded me that we needed to shut down, so I began securing the engines.

We'd been flying the EC-145 (a medium twin-turbine helicopter) for a few years, and this "cold load" policy was a special precaution on NICU flights. The isolette had to be loaded through a pair of clamshell doors at the back of the aircraft, and the tail rotor was just a step or two from the doors. Because of the potential for someone to walk into it, we didn't want the NICU team loading their equipment while the tail rotor was still turning.

As soon as I applied the rotor brake, and the blades came to a stop, Kristin began escorting the two nurses up the ramp, toward the pad. There was a security guard with them, and once they reached the helicopter, the guard helped Kristin and the nurses lift the isolette through the clamshells and into the cabin.

As soon as Kristin let me know the team was ready and their equipment was secured, I began lighting the engines again. Once we were at idle, I asked each nurse how much she weighed and plugged the numbers into my load schedule, a series of charts used to determine if we were below our maximum takeoff weight and within our center-of-gravity limitations. The numbers all checked good, so I ran the throttles up and went through my takeoff checks one more time.

"You ladies all set?" I asked.

"We're good to go," came the response from the back of the helicopter.

Kristin gave me a thumbs-up, and with that, we were off. I had no way to know in that moment that the trip from St. David's Hospital to La Grange, though it wasn't to be the final flight of my career, would be the last flight about which I have any personal recollection.

(to) take something for granted:
   1. to fail to appreciate the value of something
   2. to assume that what has been will continue to be

The half-hour trip to the hospital in La Grange was pretty much standard fare. The nurses told us the baby boy had been born prematurely, and they briefed Kristin on his condition. Trying to estimate our turnaround time at the hospital, I asked the nurses how long they would need to prep the patient for the return flight. They answered with the customary "about an hour," but from my previous experience with these estimates, I knew this meant we would likely be on the ground for a good two and a half hours. I began pondering my prospects for a late-night pizza delivery in La Grange.

When we arrived over the hospital, I discovered that another helicopter had already parked on the helipad. It was only big enough for one bird, so this meant we would have to land in the grass instead of on the concrete pad. This was no big deal. I'd done it hundreds of times on hospital transfers. It would make it harder to roll the heavy isolette in and out of the hospital, but it was nothing three nurses and a pilot couldn't handle.

Prior to setting up for our final approach, I circled the hospital once to check out the wind sock. Turning into the wind, I rolled final to a flat spot in the grass, about two hundred feet from the hospital. Once we'd landed, I shut the aircraft down and unstrapped so I could help Kristin and the NICU nurses unload the isolette.

The four of us alternately carried and rolled the unwieldy contraption, bouncing it through the thick St. Augustine grass until we reached the asphalt in front of the entrance to the emergency room. Once we were on the pavement, the wheels on the gurney became functional again; so I turned the isolette over to Kristin and the NICU team, and then I went back to finish securing the doors on the helicopter. As I turned to watch Kristin and the two NICU nurses disappear into the hospital with all of their equipment, I called the only two local pizza deliveries, but neither was open at this late hour. I began settling in for what I knew was going to be a long wait.

It was a little past one o'clock in the morning, and it was mid-October. In Central Texas, the nights are usually pleasant in the fall, and this night was no exception. There was a brilliant three-quarter moon overhead, so I decided to make myself comfortable. Lying in the cool grass, I used my survival vest for a pillow and waited. Then I waited some more, . . . followed by more waiting. This was not unusual on specialty team transfers, and I had long ago learned to accept the downtime and just make the best of it.

Eventually, Kristin called me on my cell phone and informed me that the neonatal team estimated they'd be bringing the patient out in about twenty minutes. Based on that estimate, I knew I still had about an hour to continue resting, and since I hadn't been able to score a pizza delivery, my thoughts once again turned to doughnuts. I was pretty sure Bill had gone straight down the stairs and retired to his rack, so I was resigned to the fact that there would be no doughnuts waiting for me in Austin. I began considering my other food options, figuring we would arrive back at Brackenridge at around four o'clock in the morning. The hospital cafeteria wouldn't open for another two hours, which meant I would probably have to settle for a package of stale Twinkies from the vending machine.

This was all pretty much routine after twenty years, and I had to keep reminding myself of what I always told people when they asked me if it was "cool" to be a STAR Flight pilot. I had been asked this too many times to remember, and my response was always the same.

"It sure beats the heck out of the alternative," I'd tell them, only half joking. "If I couldn't do this, I'd have to go out and get a real job."

And to tell the truth, I couldn't imagine what my life would have been like had I been forced to earn an honest living. I mean, let's face it. I was off-duty almost as many days as I worked—and calling what I did "work" was a bit of a stretch anyway. The only time I felt like I was really earning my pay was when I had to launch into marginal weather—usually on some complex, middle-of-the-night rescue mission when it was darker than the inside of a cow.

The rest of the time, I had to ease my guilty conscience by telling myself I was being paid for my skill and experience, but even that was acquired on the taxpayers' dime. I had honed my skills flying as a naval aviator for ten years, and aside from a six-month deployment to the Persian Gulf, that was a pretty good gig as well. Sure, there was a certain amount of risk that came with the job—a lot of stress as well. But all things considered, I really had no complaints . . . except for the fact that I really wanted some doughnuts.

Finally, Kristin and the neonatal nurses emerged from the hospital with our premature baby, who was resting comfortably in the warmth of the rolling incubator. I drug myself to my feet and went to meet them at the edge of the pavement. From there, the four of us repeated our earlier series of touch-and-goes across the grass to the helicopter. The total distance was about the same as from home plate to first base, and Kristin and I were doing most of the lifting. She and I were both in pretty good shape, but it was still a moderately taxing exercise.

As we positioned ourselves under the tail boom, I opened the clamshell doors, and we prepared to load the isolette into the helicopter. As was customary at this point, the two neonatal nurses became spectators as Kristin and I began to lift. I was at the front of the gurney, so it was up to me to make sure we got it high enough to slide the leading edge onto the cabin floor. That would allow the folding legs to collapse as we pushed it forward. I had done this enough times to know I was going to have to put my back into it, and normally there would have been enough room under the tail boom for me to stand fully upright with no problem. Unfortunately, in the three hours since we'd landed, the helicopter had settled in the soft ground, just enough, so that when I sprang out of my crouch and heaved up on the isolette, I no longer had enough clearance to stand straight up without hitting my . . .

The next thing I remember, I was back at Brackenridge hospital, sitting in our crew quarters—eating doughnuts. Notwithstanding my lack of confidence in him, Bill Hanson had defiantly pushed the envelope and made that unauthorized trip to the all-night bakery. And, yes—I did fly the helicopter back from La Grange that night. I just don't remember doing it. I do faintly (no pun intended) recall getting tunnel vision after hitting my head. When it happened, I knew I had to sit down—or I was going to fall down. I also remember Kristin McLain shining a flashlight in my eyes as she checked to see if I was okay. Other than that, the time between lifting the gurney in La Grange and eating doughnuts in Austin is a complete mystery to me.

According to Kristin's subsequent report, immediately after slamming my head into the tail boom, I slowly put the gurney back down, knelt on one knee momentarily, and then eventually sat down on the ground beneath the helicopter. She also said I responded appropriately when she asked me several questions during her impromptu examination. We managed to load the isolette, obviously, and Kristin said the return flight was largely uneventful. She did say she had to remind me we were flying to St. David's instead of coming back to Brackenridge, but other than that, she said the trip home was pretty much a routine flight.

Kristin must have begun to suspect something was wrong by the time we landed at Brackenridge, though. Once we had returned, she told Bill to check on me while she was completing her paperwork, and according to Bill, he knew I wasn't right because, while I was refueling the helicopter, I kept agreeing with everything he said. And it wasn't just because he'd brought me doughnuts. He said I was so agreeable that night, he could have easily talked me into selling my 400-horsepower Camaro to buy a Prius, thereby reducing my carbon footprint and doing my part to save the world. He also said I tried several times to log on to the STAR Flight computer, but I was unsuccessful because I couldn't remember my password.

It was at that point that Bill took us out of service and escorted me up to the emergency room. And according to him, I was even happy to go, which pretty much confirmed that I didn't know where I was or what I was doing. The consensus among Bill and the ER staff was that I had suffered a concussion.

For some odd reason, even though I can't remember much else, I can remember looking at the monitor at one point during the examination. I could see that my blood pressure wasn't normal, and I asked Bill about it.

"What's the deal with my blood pressure? I don't usually run that high. And why am I lying in the ER?"

"It's normal for blood pressure to run high after a traumatic injury," he said.

"What's that got to do with me?" I asked.

Bill just shook his head and said, "Don't worry about it."

By the time we got back down to the crew quarters, Mark Parcell (the chief pilot) had arrived to finish my shift. Bill and Kristin didn't think I should drive yet, so I agreed to sleep it off there at the hospital before going home. After tossing and turning for several hours, I convinced them I was okay to drive, and they let me go home.

In the days immediately following the concussion, I felt some moderate headaches and had a little trouble concentrating, but I didn't think too much about it. Within a week, however, I began experiencing slight visual anomalies. Occasionally, I would turn my head just right, and it was as if it had taken a split second for the world to catch up. In addition, I began slurring my speech, and I sometimes found myself struggling to think of words I had been using my entire life. I also lost my appetite and began dropping weight. It's hard for me to describe it, but for some reason, my food just didn't taste right to me. My neurologist would later determine that my sense of smell had been severely degraded from the concussion. That explained why I no longer enjoyed scarfing down cheeseburgers and barbecue, both of which had always been a passion of mine. I eventually lost about thirty pounds over a six-month period.

What's more, it turned out I had not only suffered a concussion that night in La Grange—I had also compressed a couple of vertebrae in my neck. They began pinching off the nerve to my left shoulder, which eventually became so painful that I couldn't raise my arm above my waist. I could barely move it away from my side, which made it difficult to perform even routine daily tasks. Then a few weeks later, the shoulder began hurting when I *didn't move it at all.* Soon after that, my twelve-year-old golden retriever died, and while trying to bury him with one arm, I tore the rotator cuff in my other shoulder. I underwent surgery to repair the torn rotator cuff, which was extremely painful and required months to rehab—leaving me to deal with two nonfunctional arms while I was struggling to overcome my post-concussion syndrome.

As a consequence, there was no such thing as exercise in my life. I could barely dress myself. I couldn't even open a bottle of beer without someone else's help, which was probably a good thing. I certainly couldn't pass a flight physical. Because the pain in my shoulders was so excruciating when I tried to lie down, I couldn't sleep in a bed. The only way I could get any rest at all was to sleep in a recliner—and that only worked for a few hours at a time. Because I had to take more and more medication just to get through the day, I gradually became dependent on the prescription pain pills the doctors were giving me.

My life was in a flat spin, and had it not been for the support I received from my wife, Nancy, and the rest of my family and friends at the time, I'm certain I would have been overwhelmed by my unfortunate circumstance. Through it all, I kept thinking I would eventually make it back to the cockpit. *I had to.* I was fifty-five years old, and I had been a pilot for my entire adult life. It was who I was, and I was good at it. Besides, I had no other marketable skills. I held a journalism degree, but the degree had only been a means to an end, a hoop I had jumped through in order to get a commission in the Navy. I had never intended to do anything but fly.

I worked with a cognitive therapist for an extended period, and during that time, I steadily began to show some promise. After about six months, I was given a series of tests, the results of which indicated some of my neurological deficits had actually subsided—though they were still significant enough to keep me out of the cockpit. Unfortunately, six months after that (a full year after the concussion), another series of tests showed only minimal additional improvement. I hadn't made any substantial gains over the ones I had seen in the first six months. I still had headaches when I tried to concentrate on a task for more than half an hour, and I still would occasionally experience short-term vertigo when I turned my head too quickly.

There *were* some good signs, however. The people who were normally around me every day noticed some improvement in my slurred speech. My ability to speak without hesitating to search for the right word was much better as well. Also, I finally managed to get some relief from the pain in my left arm and shoulder when I happened upon an innovative young surgeon at a spine clinic in Dallas. Dr. Michael Rimlawi was a miracle worker, and six months after undergoing the procedure that he himself had pioneered, the pain from my pinched nerve finally subsided. I honestly believed there was still a good chance I could someday fly again.

In the months that followed, however, there was almost no improvement in my symptoms. I was beginning to make some progress with the rehab on my shoulders, but now I was struggling to wean myself from the daily cocktail of prescription painkillers I'd been taking for the past year. I tried cutting out one medication at a time, but when that didn't work, I decided to go cold turkey. As bad as the pain from two surgeries had been, trying to free myself from the drugs I was given afterward was worse. When I stopped taking the painkillers, I was completely wired for three solid weeks. I couldn't sit still, I was constantly nervous, and a good night's sleep was totally out of the question. When I did manage to drift off, I would often find myself embroiled in a series of bizarre and disturbing dreams—most of which I can no longer remember in enough detail to describe. I just know I would wake up terrified and depressed.

Angry and frustrated with everything and everybody around me, I can only imagine what pleasant company I must have been while I was going through my withdrawal. Eventually, however, the symptoms began to lessen, and I made it through an entire month without opening the drawer where I kept the pain pills. When the chemical withdrawal had finally ended, I swore I would never allow myself to become dependent on a crutch like that again.

It had now been nearly two full years since my concussion. My neurologist ordered one more test battery for me, and this time it was administered in Houston, by a doctor who specialized in testing pilots for the medical branch of the Federal Aviation Administration. When the test was over, I really thought I had performed much better than on the previous tests. I was completely stunned and disappointed when, following the daylong exam, the doctor told me my neurological deficits were still significant—and most likely permanent. He went on to say that it was *almost certain* that I would never be able to pass the FAA's flight physical.

The words hit me like a ton of bricks. I sat in silence for several seconds, . . . and then I pressed him, asking him if he was *positive* I'd never be able to fly again. The doctor was extremely sympathetic, but then he told me I needed to come to terms with it—my career as a commercial pilot was over for good.

If the world was perfect, it wouldn't be.

—Yogi Berra (Baseball Hall-of-Famer)

So, there I was. My thirty-five-year flying career was over, and unlike most retired pilots, I didn't even have the pleasure of remembering my final flight. I had logged more than eleven thousand flight hours in almost every type of aircraft from jets to helicopters, but now the only thing I could fly was a kite.

I knew Nancy would be anxious to learn what the doctor had told me, so I called her before I started home. I'm not sure how I had expected her to react when I told her the news, but she just calmly said she loved me and told me to be careful driving home. In spite of my own optimism going into the tests, I guess she'd already had a pretty good idea ahead of time what to expect from the results. I think I even detected a note of relief in her voice. I know she felt bad for me, but I think she had been just a little bit concerned that I wasn't going to know when to throw in the towel.

Nancy and I had started dating all the way back in high school, and we'd been married for almost as long as I'd been flying. She knew me pretty well. My piloting skills weren't as sharp as they had been when I was younger, and as I'd reached my mid-fifties, the night shifts had really started to take a toll on my body. Not only was my 20/10 vision long gone, I had become more and more sleep deprived, and the stress of making life-and-death decisions was becoming just a little harder to shrug off. Toward the end, I was eating antacids like candy, and the numerous aches, pains, and gray hairs that were beginning to show up were the price I was paying for all those hours of experience in my logbook.

Still, I had convinced myself that I was able to compensate for all of the aforementioned adverse factors by applying the judgment and expertise I had gained in the process of logging those hours. When I thought about the ten years I had flown as a naval aviator, coupled with my twenty years at STAR Flight, I was convinced I had flown just about every type of mission you could fly in a helicopter.

Surely, that should count for something, I thought to myself. *I can still do this if I can just figure out a way to pass that damned flight physical!*

Maybe Nancy had a better grasp on the situation than I did, but I still thought I had a few good years left. At any rate, the decision to retire had been taken out of my hands now. The thing that bothered me most was the fact that I no longer had any control over the process, and as anyone who knows me can attest, I like to be in control.

Needless to say, I was extremely depressed as I began the three-hour drive from Houston back to Austin that night, but somewhere along a stretch of busy interstate highway, I reluctantly accepted the fact that a major part of my world was gone forever—I was never going to fly again.

On the other hand, life goes on, and in spite of the recent setbacks I'd been dealt, my life was still pretty good. I stopped at a little roadside bakery to pick up some doughnuts, and before getting back into the car, I called Nancy again to let her know everything was going to be okay.

Pulling back out onto the interstate, I had to accelerate hard to merge with the traffic. Those four hundred horses rumbling under my hood sounded really nice in the cool Texas evening. Enjoying the breeze through my open window and the sensation of the g-forces that were pinning me to my seat, I thought to myself—*At least I'm driving home in a Camaro instead of a Prius.*

2

Lee Harvey Oswald, Kosmos Spoetzl, and a German Guy Named Moses

Okay, the German guy's name wasn't really Moses. His real name was Herman "Fritz" Graebe, but his biography, *The Moses of Rovno*, likens him to the Old Testament prophet of the same name and should be required reading for anyone with a conscience.

An engineer, Fritz Graebe managed a German construction firm during World War II. His company provided services to the Railroad Administration of the Third Reich during the German occupation in Ukraine. In 1931, just like many of his peers, he had joined the Nazi party, but Fritz Graebe was no Nazi. He was a devout Lutheran, and in 1942, after witnessing the first in a series of executions of Jews near one of his work sites, Graebe knew he couldn't just stand by and do nothing. So he risked his life, and the lives of his family, by using his position as a civilian contractor to rescue scores of Jews from the Nazi killing machine.

Because his company was tasked with building infrastructure for the occupying German Army, Graebe was able to convince Nazi officials that he needed to conscript a labor force from the thousands of Jews who had been rounded up and slated for execution. Often fabricating nonexistent construction projects, he would requisition Jewish workers and then shelter them from the Nazis until he could smuggle them to safety. The Israeli government would later recognize Fritz Graebe for his bravery by planting a tree in his honor along the Avenue of the Righteous, a memorial in Jerusalem that is dedicated to all those who helped to save Jews from the Holocaust.

In addition to rescuing Jews in Ukraine, Graebe became the only German citizen to voluntarily testify against the Nazis at the Nuremberg Trials following the war. He came forward even though he knew he would be scorned for doing so by many of his fellow countrymen, who wound up ostracizing Graebe and his family. The postwar German economy was already in shambles, and resentment for his prosecution testimony at Nuremberg made it even harder for Graebe to earn a living.

After receiving numerous death threats, Fritz Graebe packed up his family, including his son, Frederick, and immigrated to the United States in 1948. He started his own construction company in San Francisco, where he was highly respected and active in the local German-American community until his death in 1985. Fritz Graebe's son, Frederick, also became an engineer and, after that, an attorney for the California Department of Transportation, where he was a public servant for more than forty years. Frederick Graebe had a son named Harold, who became a naval aviator.

Harold Graebe, the grandson of Herman "Fritz" Graebe, served with me in the Persian Gulf during Operation Earnest Will in 1987. He was, and still is, the best pilot with whom I've ever flown.

Kosmos Spoetzl was born in Bavaria, Germany, in 1837. By all accounts, he liked good food, good cigars, and good beer, not necessarily in that order. He was a brewmaster and eventually became the sole proprietor of the Spoetzl Brewery in Shiner, Texas, a Czech and German community located about ninety miles southeast of Austin.

The brewery was founded in 1909 by a group of local residents who hired Spoetzl to come to Texas and produce the old-world beer they had enjoyed in their countries of origin. Several years later, he purchased the brewery, and the rest, as they say, is history—at least it is to those of us who live in Texas and enjoy wetting our whistles from time to time.

Today, you can find Shiner Bock in just about every state in the nation. However, even though Bock is the most prolific of all the Shiner beers, it was only a seasonal brew until 1973. Not long after the dark lager became available year-round, it became a favorite among young, affluent professionals in Texas, many of whom had migrated from other states. The brewery prospered greatly from Shiner Bock's popularity, and today it's the oldest locally owned brewery in Texas.

The first (and still the best) Shiner beer brewed in 1909, however, was Shiner Premium, a Bohemian-style golden lager. Subsequently known as Texas Special and Texas Export, it was also embarrassingly relabeled Shiner Blonde a few years back—but, mercifully, that didn't last long, and now it's once again being sold under the original and proper "Premium" label. Not nearly as ubiquitous as Shiner Bock, Shiner Premium is still the preferred beverage of many native Texans, myself included. Now, don't get me wrong. Shiner Bock is an excellent brew. It's certainly not a mundane corporate beer, and I don't mean to imply that it's in a category with Lone Star and other Texas *tourist* beers. I enjoy a Bock as much as the next guy, but the point here is that, to those who proudly think of themselves as true Texas-beer purists, there's absolutely nothing better than an ice-cold Shiner Premium, especially when served with barbecue and other fine Texas cuisines. Heck, at least one Texas-beer purist has even been known to enjoy a few ice-cold Shiner Premiums while writing his book.

Sadly, when it comes to fine Texas libations, the average bar patron doesn't share my discerning pallet, so he routinely elects to follow the herd and blindly orders whatever beer happens to be trending within his social circle. Therefore, most restaurants and bars cater to the masses by offering Shiner Bock instead of Shiner Premium. There is still one iconic, true-to-Texas establishment, however, where Shiner Premium is still proudly served. It's the legendary Texas Chili Parlor in downtown Austin, Texas.

Stay with me here. I promise I'll get to Lee Harvey Oswald.

Naval Air Station Pensacola, Florida, is known as the "cradle of naval aviation." Home to the Naval Aviation Schools Command, it's where every prospective naval aviator begins his flight training. As a newly commissioned ensign, I reported to Pensacola for Aviation Indoctrination, the first phase of that training, in the spring of 1983. Having attended the University of Texas at Austin on a Navy ROTC scholarship, I had received my commission upon graduating from that esteemed institution ten months earlier.

That same year, Harold Graebe graduated from the University of California at Davis and was accepted into the United States Navy's Aviation Officer Candidate School, also in Pensacola. AOCS, as it was known, was a program whereby civilian college graduates who wanted to become naval aviators could earn a commission while concurrently completing the indoctrination portion of their flight training.

Once they had earned their ensign bars, the AOCS graduates were integrated with the already-commissioned student naval aviators to begin primary flight training in one of four fixed-wing training squadrons. Three of the squadrons were based at NAS Whiting Field, northeast of Pensacola, and the other was at NAS Corpus Christi, in Texas. Harold Graebe was assigned to Whiting Field, where he completed his primary and intermediate fixed-wing training, as well as his advanced helicopter training. In contrast, my journey through flight school followed a much more oblique route.

After I had successfully completed the Aviation Indoctrination course (despite nearly drowning during the requisite survival-swim training), my wife Nancy and I packed our bags and headed back to our home state of Texas. I had been assigned to the primary training squadron in Corpus Christi, where I began my Navy flying career in the North American T-28 *Trojan*. The T-28 was a beast of an airplane. Powered by a radial engine packing 1,425 horsepower, it represented the last vestige of a bygone era, a time when propeller-driven airplanes were big, fast, and—best of all—

loud. It was truly a man's airplane, complete with a supercharger, electric cowl flaps, and a hydraulically actuated speed brake. Equipped with hard points on the wings to accommodate various armament packages, the T-28 was more than just a trainer. It was routinely used as a close-air-support and attack aircraft by the U.S. and South Vietnamese Air Forces during the Vietnam War, and over its many years of service, the *Trojan* was flown in combat by various other air forces around the world.

The Navy began flying T-28s in the early 1950s, and by 1983, they had almost been completely replaced by the Beechcraft T-34C *Turbo Mentor*. There were just a handful of T-28s still flying, and they belonged to VT-27, based at NAS Corpus Christi.

Three weeks before my class graduated from Indoc in Pensacola, we were informed that the Navy needed to fill three final T-28 slots in Texas, which surprised us because we had been told we would all be training in T-34s. The lieutenant who was briefing us said that the three who volunteered for these slots would be the last naval aviators to train in the T-28 before it was permanently retired from the Navy's inventory. My hand shot up even before he had finished talking. I remembered having been disappointed a year earlier upon learning that I wouldn't be starting flight school until ten months after receiving my commission because, at the time, I was sure all the T-28s would be gone by the time I began my primary flight training. But now I was catching a huge break. I was going to be in the final class of student naval aviators to train in an airplane that embodied a significant piece of naval aviation history.

My primary flight training at NAS Corpus Christi was one of the most exciting times in my life. Nancy and I lived on base, and every day, from 08:00 to the end of flight operations, you could hear the ever-present reverberation from the parade of T-28s, with their massive radial engines, roaring over our house. I don't think Nancy enjoyed it nearly as much as I did, but even though it was almost impossible to sleep until the last plane had landed (usually around midnight), she never complained.

Bigger, louder, and faster, the T-28 was much more complex than the T-34 that replaced it. This made it both more exhilarating and more challenging to fly. In addition, the T-28 required a longer and more arduous training syllabus, and I believe we became better pilots for having flown it. After eight months of intensive training, my classmates and I graduated from primary, and I was fortunate enough to finish at the top of that final T-28 class.

There's little doubt my previous flight experience (I had flown several hundred hours as a civilian pilot) had been an advantage, but there was also no denying that I was more focused than the average student naval aviator. I had made up my mind that I was going to earn those gold wings, and I wasn't about to let anything distract me until the task was completed. At any rate, I was one of the lucky few who had the privilege of retiring a historic aircraft to the boneyard when we flew the last of the T-28s to be mothballed at Davis-Monthan Air Force Base in Tucson, Arizona.

The next time I would touch a T-28 wasn't until many years later. Nancy and I were driving past a transportation museum in North Texas, not far from Keller, the town where we had first met while still in high school. One of the airplanes on display in front of the museum was a T-28B *Trojan* with VT-27 markings on it. We had our kids in the car with us, and I decided to turn around so we could stop and let them see it up close. Lee Ann, our daughter, was about eight years old, and James, her younger brother, would have been about five at the time. To them, it was just another airplane, and an old, "beat-up" airplane at that. The paint was faded and peeling, and the tires on the landing gear main-mounts were completely flat. To me, however, it was a piece of history, not to mention an important part of my life.

I wrote down the serial number, and later, when I checked it against my logbook, it was indeed one of the planes I had flown during my training at NAS Corpus Christi. Even though I was only in my mid-thirties at the time, it actually made me feel a little old to discover that a plane I had flown as a primary trainer was now a museum piece. Some twenty years after that discovery, I would find another T-28 in which I had trained, and this one was still flying. It was part of an impressive collection of historic aircraft at the Cavanaugh Flight Museum in Addison, Texas. That was several months after my concussion, however, and even though my old primary trainer was still in good flying condition—yours truly was not.

From NAS Corpus Christi, I was assigned to intermediate jet training at VT-23, based at NAS Kingsville, Texas. There I flew the T-2C Buckeye, an old straight-winged, twin-engine jet from the Korean War era. Like the T-28, it was manufactured by North American Aviation, and just like the T-28, it was soon to be retired from the Navy's inventory. Surprisingly, I found the T-2 much easier to fly than the T-28, and just as in primary, I was doing pretty well and making top grades.

There was just one problem. It was a phenomenon known as "departure from controlled flight." A major part of the T-2 syllabus was the *air combat maneuvering* phase, and one of the things we had to learn during that phase was how to recover from accelerated stalls. When an airplane stalls at very high airspeeds (as it has the potential to do during extreme, high-g maneuvers), it doesn't just stop flying—it becomes uncontrollable. The pilot can't simply lower the nose as he would to recover from a normal *one-g* stall because, once it has *departed* (jet jock vernacular), the aircraft is tumbling end over end. Normal control inputs become useless, and the pilot is just along for the ride.

Eventually, usually after a half minute or so, the cartwheeling plane will settle into a steady-state spin from which the pilot can effect a recovery. During those thirty seconds, however, it's like being inside one of those machines they use to shake paint cans in the hardware store. Even though I was always able to recover and continue flying after the cartwheeling maneuvers, I usually came back with my most recent meal in a plastic bag.

For someone who was determined to become a naval aviator, this was extremely demoralizing and discouraging. I was set on making this my career, and this bout with motion sickness posed a serious threat to those plans. I couldn't understand why I had never experienced this problem before now. During my primary training, I had performed plenty of aerobatic maneuvers in the T-28, and motion sickness had never been an issue. Later on, when I deployed on cruisers and frigates, I never once got seasick—even in sea states that left half the crew incapacitated. As for my air sickness, positive g-forces (the kind normally experienced during aerobatic maneuvers) weren't the problem. That's why I hadn't gotten sick while flying aerobatics in primary. It was those *damned* oscillations that were making me sick during the accelerated-stall recoveries.

After some serious soul-searching, I decided to request a transfer from jets to helicopters. I understood enough about helicopters to know that pulling negative "g"s was not part of the program. My commanding officer, however, was convinced I could overcome my motion sickness. He even wanted to send me back to Pensacola to attend a supplemental training course affectionately known as "puke school." That's right—the object of the training was to *make* you puke!

A bunch of Navy flight surgeons had developed the course for guys just like me. The idea was to keep subjecting the student to vomit-inducing carnival rides until he eventually became inoculated against motion sickness. Call me crazy, but the idea of letting a bunch of pointy-headed guys in white lab coats spin me around until I barfed my guts out repeatedly didn't sound all that appealing to me. After a lot of discussion, I was able to convince my CO that I really *did* want to fly helicopters instead of jets, and he agreed to let me skip puke school. He endorsed my request for a transfer and forwarded it to the Chief of Naval Aviation Training for final approval.

Several of the VT-23 flight instructors, and a few of my classmates as well, were extremely disappointed in me for deserting the jet community just so I could perform the menial task of flying a helicopter. You have to understand that, to some guys who flew tactical jets, helicopter pilots were second-class aviators, not worthy of cool call signs like "Viper" and "Iceman" (mine was "Lobo," by the way, which I think was pretty cool). I often joked that, in the mind of a typical jet driver, the lead role in a *Top Gun* movie about Navy helicopter pilots would be played by the Pee-wee Herman character instead of Tom Cruise. Never mind the fact that, in the Navy, each and every helicopter pilot was also qualified to fly airplanes.

On the flip side, the average restricted (fixed-wing-only) aviator, left to his own devices, would almost certainly end his day as a smoking hole in the ground if you turned him loose in a helicopter. And it goes without saying that once our high and mighty jet driver has ejected over the water or behind enemy lines, the helicopter pilot suddenly becomes his best pal.

This is a broad generalization, of course. Most of the parochial bias that existed within the various naval aviation communities was benign in nature. Later on, after I had earned my wings, I was even guilty of it myself to a degree. I would sometimes antagonize my buddies who wound up flying maritime patrol aircraft, mainly P-3 *Orions*, by asking them to explain how they justified the anchors on their wings. Maritime patrol pilots were shore-based, and as such, they were the only naval aviators who didn't operate from ships and never spent a day at sea. Naturally, those of us who were in sea-going billets were quick to remind them just how easy and boring their missions were in comparison to ours. With few exceptions, however, the back-and-forth between the different communities was always tongue-in-cheek.

I continued treading water at VT-23 until my pipeline transfer was approved, at which time I was assigned to NAS Whiting Field, back in Florida. This is where my new classmates and I received the advanced training required to make us helicopter pilots.

I vividly remember my first flight in the Bell TH-57. Never having flown (even as a passenger) in a helicopter before that day, it was all brand new to me. Flying a helicopter in cruise flight is really not much different from flying an airplane, so it was no big deal when my instructor initially let me take the controls after we were safely cruising at altitude. Even though it had been several months since I had flown a T-2 back in Kingsville, I had very little trouble maintaining my altitude, heading, and airspeed as we flew west, along one of the training routes from Whiting Field.

About ten minutes into the flight, my instructor took the controls back from me, and we made our approach to Spencer Field, one of several auxiliary training sites north of Pensacola.

Once we had claimed our spot on the field, he simultaneously demonstrated and explained the control inputs required to hover. He made it look pretty easy, and I was anxious to try it for myself. After all, it's the ability to hover that actually defines a helicopter pilot and separates him from his fixed-wing brethren.

The drill was to make the student responsible for just one control axis at a time. He started by letting me take the pedals. All I had to worry about was keeping the nose of the helicopter pointed into the wind, which was really not all that hard. Then I had to make left and right pedal turns, rotating the nose of the helicopter three hundred and sixty degrees while the instructor maintained our hover over a fixed point on the ground. This was a bit of a challenge at first, especially halfway through the turn, at which point the wind hit the tail like a weathervane and rapidly increased our turn rate.

After completing several turns, however, I began to get the hang of it, so the instructor took the pedals back and gave me control of the collective. Now all I had to do was maintain a constant altitude. If the helicopter began settling, I was instructed to raise the collective (ever so slightly) to arrest the descent, then climb just enough to regain the original hover altitude—at which point I had to lower the collective again to avoid climbing too high. The inputs to accomplish this were extremely subtle, and this part of the exercise was much more difficult than operating the pedals. Just as I had with the pedals, though, I eventually settled down and was able to maintain a reasonably constant altitude as we hovered in place.

This is where it really began to get interesting. I was relieved of the collective and given command of the cyclic. Now my job was to keep the helicopter over the fixed point on the ground. This didn't go well at all initially, as even the slightest movement of the cyclic sent us careening away from our spot and into the next zip code. To make matters worse, I found I couldn't just hold the cyclic in one position and forget about it. Every time the wind would gust, even just a little, I had to move the cyclic in the opposite direction to compensate for it. Then, when the gust subsided, I had to move it back or off we'd go again, this time in the opposite direction. The instructor had to take the cyclic away from me close to a dozen times before I was finally able to keep the aircraft within the confines of our working area on the field.

This is how the Navy taught us to hover a helicopter, one control at a time—then two—and finally, all three together. It's a hard thing to impart, much like trying to teach your kid how to ride a bicycle. Because it's a seat-of-the-pants thing, there's no good way to explain it. The student has to learn by feel. One day your kid falls over when you let go of the bike—the next day he pedals off down the street. It doesn't matter how smart your kid is, or how many books he reads on how to ride a bike—the only way he learns to stay upright is by trial and error. It's the same for a pilot learning how to hover. People often ask me, "Which is harder to fly, the airplane or the helicopter?" All I know is that even though I had already logged several hundred hours in airplanes when I reported to Whiting Field, learning to hover a helicopter was, without a doubt, the most challenging phase of all my Navy flight training.

Once we could actually take off, hover, and land again without killing ourselves, our training had just begun. The transition from fixed-wing pilot to helicopter pilot was a six-month process. The next major hurdle, after learning to hover, was learning how to safely perform engine-out landings, or autorotations. These had to be mastered both from a hover and from cruise flight at altitude, and they were undoubtedly the riskiest maneuvers we performed during our training.

After that, we learned how to takeoff, fly, and land the helicopter without any visual reference, using only the cockpit instruments. We also practiced "slinging" external loads and landing in confined areas. We learned how to navigate using only a geological survey chart—and we did it while flying in high-speed formation with a second helicopter, at an altitude barely above the tree tops. We also learned how to conduct combat search-and-rescue operations. This included making high-speed tactical approaches, designed to minimize our exposure to ground fire while entering hot landing zones. And finally, we learned to operate to and from ships under way at sea. After all, we weren't just training to become helicopter pilots—we were training to become naval aviators, the *best* helicopter pilots in the world.

A word of warning here: Ask a naval aviator what it is that makes him better than other pilots, and you have unwittingly granted him a forum that he will gladly use to deliver a lengthy, self-serving dissertation designed to convince everyone who happens to be within earshot that he is superior to those less fortunate pilots who learned their piloting skills in other branches of the military.

Furthermore, if said dissertation happens to take place inside a bar, it means that any anecdotal evidence the naval aviator offers in the form of sea stories (especially those beginning with the aforementioned phrase, "there I was") need only retain a modicum of truth as it relates to the original facts surrounding the event being described. Having met that requirement, the gold-winged narrator is subsequently free to embellish his recounting of the event to whatever extent he deems necessary in order to convince his fellow bar patrons that he possesses flying abilities far superior to those of ordinary pilots.

If the naval aviator in question happens to be a Texan, then the *modicum-of-truth* requirement is automatically waived. If the naval aviator also happens to be a Marine, all bets are off, and there is no longer a reasonable expectation that the story being told is based on any sort of real facts. Fortunately, my integrity has only been compromised by the first of these aggravating factors. Ergo, as the author of these chronicles, I am still constrained by the *reasonable-expectation-of-a-factual-basis* rule. This affords you, the reader, the right to accept *most* of the narration in this book as a truthful representation of actual events—occasionally enhanced with a small measure of overstatement.

Also, in defense of the naval aviator, I would like to point out that his inflated opinion of himself is not entirely his fault, and he should therefore be granted some leeway for thinking the way he does. His hyperbolic sense of self-worth is not the result of some inherent character flaw. The reality is, he just can't help himself. It's not as though he doesn't possess the decorum necessary to walk into a room without swaggering. The problem lies in the fact that he's entirely unaware he's doing it. When others think he's being arrogant, he thinks he's just being candid.

In order to comprehend it, you need to know that as soon as every young and aspiring student naval aviator passes through the gate at NAS Pensacola, he is immediately subjected to a highly effective dose of indoctrination. From day one, his instructors begin preaching to him, telling him he's on a quest to join an ultra-elite group of combat pilots. As he hears this mantra repeated throughout twelve to eighteen months of highly intense training, he either buys into it, or he washes out. So if he's fortunate enough to complete his quest, it's little wonder when, as he surveys the societal command structure, he fancies himself only one or two grades in rank below the Deity—especially after he's been presented with a multi-million-dollar aircraft, entrusted to him by a grateful nation of taxpaying citizens. Why do you think the Navy's flight demonstration team is called the Blue Angels?

So, it's within this context that I say the following: When my training was finally over, and Nancy pinned those gold wings on my chest, it was one of the proudest days of my life. At last, it was my turn to be that annoying guy in the leather flight jacket, the one who bores people to tears at parties. I was now the larger-than-life O'club warrior, the protagonist to whom anyone who is not already engaged in conversation represents a target of opportunity, just waiting for me to regale them with inflated accounts of my aeronautical expertise—in a pleasant and unassuming way, of course.

Now, getting back to Harold Graebe (remember him?), he was the grandson of Fritz Graebe, the guy who rescued all those Jews from the Nazis. Harold had already completed his Navy flight training several months ahead of me. After receiving his wings, he was assigned to HSL-43, based at NAS North Island in San Diego, California. HSL stood for Helicopter Antisubmarine Squadron Light, and HSL-43 was a brand new squadron, sporting brand new helicopters—Sikorsky SH-60B *Seahawks*.

The *Seahawk* was the Navy's newly acquired variant of the Army's UH-60 *Blackhawk*. This was a choice billet for a "nugget" (the term used to describe recently winged naval aviators), and only the top few from each graduating class were chosen for it. Harold had been chosen because, as I stated earlier, he was the best pilot with whom I've ever flown.

As much as I hate to admit it, Harold's flying skills were even better than mine—and mine were pretty good. I had the grades to prove it, too (remember what I just told you about bragging in a pleasant and unassuming way?). Of the more than seven hundred naval aviators who received their wings the same year I did, I managed to graduate in the top one percent. That means I ranked somewhere between first and seventh out of seven hundred. I don't have any way to know exactly. My citation just said I was in the top one percent. The point I'm trying to make is that, at the risk of sounding arrogant, I was better than most young naval aviators—and naval aviators, as a group, are already pretty good (not to mention . . . arrogant?).

Still, I can't deny that, between the two of us, Harold was the better pilot. And I'm not just saying that to be magnanimous. Trust me, it would bring me far greater pleasure if I could say that *yours truly* was the greatest helicopter pilot the Navy had ever produced, but I'm trying to write an honest book here (remember, I'm bound by that *modicum-of-truth* thing). I later learned that Harold had graduated in the top one percent of naval aviators that year, just as I had. With that in mind, it means that, out of those seven hundred naval aviators I mentioned in the previous paragraph, I must have actually been ranked somewhere between *second* and seventh. I'm quite sure Harold Graebe was at least one place ahead of me. At any rate, Harold and I both still had a lot of lessons to learn about flying helicopters in Uncle Sam's Navy. We would soon be learning many of those lessons while flying together as wet-behind-the-ears *Seahawk* drivers.

Several months after Harold Graebe had received his orders to HSL-43, I was excited to learn that I was receiving orders there as well. When I finally joined him in San Diego, which is where we first met, life there for nuggets was about as good as it gets. Because ours was the first West Coast fleet squadron to begin flying the new SH-60s, we had all new state-of-the-art facilities, including a brand new hangar in which to house our freshly minted helicopters.

What made it even better was the fact that the Navy brass had managed to procure a complete complement of aircraft for our new squadron long before we had enough pilots on board to fly them. This meant that every single aviator assigned to HSL-43, from the saltiest veteran to the shiniest new nugget, had to pull his weight by logging as much flight time as he possibly could. This created a rare opportunity for the junior pilots, including the nuggets, to fly with one another.

Because there weren't enough qualified HACs (Helicopter Aircraft Commanders) to go around, nuggets were allowed to fly together as pilot and copilot. Normally you had to qualify as a HAC (pronounced *hack*) before being allowed to sign for an aircraft, which meant that newly qualified copilots like Harold Graebe and I could only fly with the more senior pilots in the squadron. But because we had too many helicopters and not enough crews to fly them, our commanding officer decided to waive this requirement. As a result, Harold and I wound up flying together quite often, and by the time we both became HACs, we had forged a pretty good friendship, largely out of mutual respect for one another's flying skills.

Later, we were both assigned to the aviation detachment aboard the USS *Valley Forge*, which was one of the Navy's newest guided missile cruisers. In 1987, when we deployed to the Persian Gulf in support of Operation Earnest Will, I went out of my way to request Harold as my copilot. Of course, if Harold were telling this story, I'm sure he'd tell you that, instead of asking him to be *my* copilot, I asked to be *his* copilot. No matter. We were both HACs by then, each having logged more than a thousand hours in the SH-60, so we divided the pilot-in-command duties evenly, each of us flying as aircraft commander and copilot on alternate missions.

The salient point here is that I wanted to fly with Harold because I believed it would significantly increase my survivability while flying in a hazardous environment. Another reason I chose to fly with Harold was because it wasn't just the skill he displayed as a naval aviator that had earned my respect. He also possessed a great deal of character and integrity. It was the same sort of moral strength and courage his grandfather had displayed when he risked his life standing up to the Nazis during the Second World War—which brings me back to the title of this chapter. I still haven't explained the connection between Fritz Graebe, Kosmos Spoetzl (the Shiner Beer guy), and Lee Harvey Oswald (I *promised* you I'd get to him).

In November of 1963, I was six years old. I didn't start elementary school until the following year, so I was at home, watching television with my mother, the day it happened. The Kennedy assassination is the earliest recollection I have of my childhood, and therefore it represents a chronological benchmark for how I view historical events.

Because I wasn't there to witness them, all I know about events that occurred prior to the assassination is what I learn through books and other media. *The Moses of Rovno*, written by Douglas K. Huneke, is a good example. By reading the book, I learned about one of the many historically significant, yet relatively unknown, individuals who greatly contributed to humankind by standing up against the Nazis during the Holocaust. In addition to what I learned about Fritz Graebe by reading his biography, my decades-long friendship with his grandson, Harold Graebe, affords me a personal connection to the man about whom the book was written.

In a sense, I have a nebulous personal connection to Lee Harvey Oswald as well. Everyone knows who Oswald was, and most people know he had a Russian-born wife named Marina. What many people may not know is that he also had two daughters.

When I first saw Rachel, she was working her way through nursing school as a waitress at one of my favorite hangouts, the Texas Chili Parlor, in Austin, Texas (the same place I mentioned way back at the beginning of the chapter). As I told you earlier, the Chili Parlor is the place to go if you're looking for Shiner Premium, and I had been a regular at the Chili Parlor dating back to the late 1970s, when I was a student at the University of Texas. As a naval aviator in the late eighties and early nineties, I made numerous cross-country flights to Bergstrom Air Force Base, which later became the Austin-Bergstrom International Airport.

I never flew to Austin without dropping in at the Parlor; and at first, I only knew her as Rachel, the waitress who was a nursing student at the University of Texas. Not that I would have expected her to, of course, but she never told me who her father was. Still, I knew she looked vaguely familiar (Rachel bore a striking resemblance to her mother), and when I asked her one day if Austin was her home, she said she had grown up in Rockwall (a northeast Dallas suburb). Then one night, I heard her introduce herself as Rachel Porter to someone at the bar. It was then that I realized who she was.

I was a good enough student of history to know that after Marina Oswald's husband had been killed by Jack Ruby, she married a man named Kenneth Porter, who lived in Rockwall, Texas. This was before the Internet age, so I actually had to do some research at the UT campus library to satisfy myself that my supposition about Rachel was correct. That's where I confirmed that Lee Harvey Oswald was, in fact, survived by Marina (Porter) and two daughters, the youngest of whom was named Rachel.

I never let Rachel know that I was aware of her father's identity, although I'm not sure why. It's not as if it was a closely guarded secret. Later, while still working at the Chili Parlor, she even granted the media a couple of interviews on the subject. I guess I just felt a little guilty that I had confirmed it "behind her back" instead of just asking her about it. At any rate, I admired the way she had managed to live a normal life despite her father's notoriety. I also respected her for working her way through college. My degree had been taxpayer-funded, courtesy of the United States Navy. Rachel, on the other hand, was earning hers the hard way.

Connecting the Dots

So what is Lee Harvey Oswald's connection to Fritz Graebe (Harold Graebe's grandfather) and Kosmos Spoetzl, the Bavarian brewmaster? Although the name Lee Harvey Oswald is historically noteworthy, the nexus actually has nothing to do with the man himself, or the fact that he's remembered for assassinating President Kennedy. Instead, the connection lies in the fact that his daughter, while waiting tables at the Texas Chili Parlor, unwittingly planted a seed that would eventually lead me to make a life-changing career decision.

It happened one night while she was working the table where Harold Graebe and I happened to be eating dinner at the Texas Chili Parlor. I had already introduced Harold to the Chili Parlor on one of our previous trips to Austin because, as you already know (if you've been keeping up at all), it was one of the few places where you could enjoy Shiner Premium while dining.

Well, we had *enjoyed* several rounds of said Shiner Premium when we heard the unmistakable rumble of a good-sized helicopter passing low over the bar. Naturally, we were curious, so I asked Rachel if this was a common occurrence. She told me it was STAR Flight, the EMS helicopter based at Brackenridge Hospital, just a few blocks away.

This was surprising to me, given that when I had graduated from college back in 1982, Austin didn't have an EMS helicopter. I asked Rachel how long there had been a helicopter flying out of Brackenridge, and she said it had been around for several years.

For the rest of that evening, the prospect of a post-Navy flying career was the main topic of discussion at our table. Harold and I were both nearing the end of our hitches in the Navy, and Rachel suggested that, while I was in town, I should go talk to the people at STAR Flight about flying for them.

I decided I would do just that, and the next day, I made my way over to Brackenridge Hospital. There I discovered a Bell 412 (basically a twin-engine *Huey* with four main-rotor blades) sitting on the pad next to the emergency entrance. It was a good-looking bird—sitting on high skids and sporting a nice paint scheme of blue and red stripes over a shiny white base coat. On the sides, the words "STAR Flight" were painted in blue italics on each of the two sliding cargo doors.

There was a security guard standing near the pad, and I wanted to take a look inside the cockpit, so I asked him where I could find the on-duty crew. He took me down a set of stairs, which led to a parking garage, located directly under the pad. Just to the right, at the base of the stairs, was the door to the flight crew's living quarters. I think the guard was a little suspicious of me, because he hung around while I knocked on the door. When the door opened, I introduced myself and explained why I was there. At this point, the security guard seemed satisfied that I was not a threat, so he left me there with the person who had opened the door.

Standing there in front of me was a guy wearing a blue flight suit. He introduced himself as Jim Allday and told me he was a paramedic on the helicopter. I asked him if I could talk to the pilot, and he said the pilot was taking a nap—which, naturally, got my attention right away. The fact that the pilot was taking a nap during his shift made the idea of becoming an EMS pilot all the more intriguing to me. Even though the pilot was asleep, Jim offered to take me up to the pad and show me the helicopter, so I took him up on it.

I followed Jim back up the steps to the helipad, and as he was giving me the grand tour, he seemed more than happy to answer all my questions. I later learned that this kind of dog and pony show was right in Jim Allday's wheelhouse. He loved to talk, and he especially loved to talk about STAR Flight. The fact that he could do it in a passionate and articulate fashion made him the perfect ambassador for the program.

I asked Jim if I could sit in the cockpit, and he told me to climb in. It was extremely well equipped and featured a fully rated IFR instrument package, which meant the aircraft was capable of being flown under IFR (instrument flight rules) in bad weather. I didn't know it at the time, but this all-weather capability would one day prove useful to me.

Jim also showed me how the stretchers were loaded into the aircraft by means of swiveling contraptions that resembled lazy Susans, which were mounted to the deck, just inside each of the cargo doors. Then he showed me something called a reconfiguration kit, which could be used to erect the necessary framework for stacking two additional stretchers above each primary stretcher. This meant that, if necessary, the aircraft was capable of transporting up to six patients. Jim even took the time to educate me (in more detail than I actually needed) about each piece of medical equipment that was carried aboard the aircraft.

This was all very impressive, but when I asked where I should send my résumé, it became a little confusing. Jim explained that the crew consisted of a pilot, who worked for Travis County, a paramedic, who worked for the City of Austin EMS Department, and a nurse, who worked for Brackenridge Hospital. The helicopter was owned by the county. I asked him if this arrangement wherein everyone reported to a different boss ever caused any problems between the crew members. He just grinned and didn't answer, which led me to believe I had stumbled onto a sensitive subject.

I thanked Jim for showing me the helicopter and answering all my questions, and he shook my hand and wished me well. I didn't know it at the time, but Jim Allday and I would eventually become longstanding friends and colleagues. Over the course of what would become a two-decades-long association, the two of us were dispatched on a lot of missions together; and as fate would have it, Jim would wind up flying with me on a night neither of us would ever forget.

The Butterfly Effect

It's funny how the script unfolds sometimes. Each man who is put on this earth has the potential to touch the life of every other man, even those with whom he never crosses paths. Whether you call it fate or divine providence, it's hard to deny that what happens in the lives of people we don't even know can, and sometimes does, affect each of us personally. Most of the time, we don't even understand how the seemingly insignificant things that happen in our own lives can profoundly affect not only *our* destiny, but the destinies of those around us as well.

Had it not been for my embarrassing motion sickness problem, I would have continued flying jets instead of becoming a helicopter pilot. Had I not become a helicopter pilot, I would never have met Harold Graebe. For that matter, Harold and I would never have met if his grandfather had not been ostracized and forced to move from Germany to the United States. If I had never met Harold Graebe, and if it were not for the fact that the Texas Chili Parlor is one of the few places available to enjoy Shiner Premium beer, he and I would not have been there the night Rachel Porter encouraged me to stop by the STAR Flight crew quarters. Had I not followed her suggestion, there's a pretty good chance I would never have pursued a career as an EMS pilot, in which case some of the people who were subsequently rescued while I was flying for STAR Flight might not be alive today.

So it follows that my life *and* the lives of many others have all been affected by a common link—the curious connection between an accused presidential assassin, a man who is best known for brewing high-quality Bavarian beer, and the man known to Holocaust survivors and their relatives as the Moses of Rovno.

Go figure.

Our wedding at First Baptist Church of Keller, Texas (September 11, 1982)—Since the terrorist attacks on the World Trade Center and the Pentagon in 2001, we've known what it was like for those couples who, because they were married on "just another" December day prior to the attack on Pearl Harbor in 1941, had to celebrate their subsequent anniversaries on the 7th of December, the date that FDR famously declared "will live in infamy."

Nancy's senior picture at Keller High School (1975)—
It's easy to see why she garnered my attention.

Nancy posing next to my "ensign mobile" (a 1982 Pontiac Firebird) at NAS Corpus Christi, Texas (1983)—Because they suddenly had steady incomes at their disposal, it was customary for newly commissioned ensigns to run out and buy a new car as soon as they graduated.

Yours truly (left) with two of my primary flight training classmates at VT-27.

Standing in front of a T-28 *Trojan* on the NAS Corpus Christi flight line—During the summer and fall of 1983, we were the last class to fly this historic aircraft.

Receiving my naval aviator designation from the commander of Training Air Wing FIVE at NAS Whiting Field in Florida—Minutes later, Nancy would pin the coveted wings of gold to my uniform (May 14, 1985).

A brand new SH-60B *Seahawk* over the San Diego Bay Bridge—When Harold Graebe and I reported to HSL-43 in 1985, this was a highly sought-after assignment for Navy helicopter pilots, especially for a couple of "nuggets."
Photo courtesy of Sikorsky Aircraft Corporation

Relaxing next to the cockpit of my *Seahawk* at an air show in Los Angeles, California.

Nancy gives me a big welcome-home hug (center of photo) upon our return to San Diego from the Persian Gulf in October, 1987. Harold Graebe can barely be seen (back left), as he's greeted by his wife, Ingrid. We flew to NAS North Island while our ship was still fifty miles offshore, thereby shortening our deployment by several precious hours. *Photograph courtesy of Gerry Spencer*

Nancy and I attend a squadron mate's wedding in November, 1987—If you look carefully, you'll notice our daughter, Lee Ann, is also in the photo (she was born just ten days later at the Balboa Naval Hospital in San Diego).

Harold and Ingrid Graebe at the same wedding—Because of the bond forged between our families while serving together at HSL 43, Harold and Ingrid are still our good friends to this day.

With Nancy at a squadron "fifties party"—Although it was fitting a little tighter by 1987, the letter jacket is the one I actually wore at Keller High School back in the mid-seventies. The poodle skirt and saddle shoes, even though she wore them well, were not part of Nancy's high school wardrobe.

During my flight instructor tour at NAS Whiting Field, one of my duties was to fly Santa in for the squadron Christmas party. That's our daughter, Lee Ann, on Santa's lap and our son, James (a bit camera shy), in my arms.

Three decades after I had flown the last of the Navy's T-28s, I found one in Texas that was still flying. *Photos courtesy of Mark Parcell.*

3

A Chronic Case of Aerodynamically Induced Paranoia

The great philosopher Forrest Gump once said, "What's normal anyways?" Well, in the world of aviation, it's generally accepted that helicopter pilots are, for lack of a better term, "different." For this to make sense, you have to understand that the helicopter, when compared to an airplane, is a "different" kind of machine. Some differences are subtle, while others are quite stark. Each is a heavier-than-air aircraft, and each relies on airfoils (wings) to generate the lift required to overcome gravity. The airplane's "fixed" wings can only produce lift from the relative wind that flows as the plane moves forward through the surrounding air mass. Helicopter wings, because they are constantly in motion, can generate the relative wind necessary to create lift even when the helicopter is stationary. This is what enables a helicopter to hover in place.

This unique ability to hover is the primary thing that distinguishes the helicopter from every fixed-wing airplane dating back to December 17, 1903. That's when Orville Wright, with his brother, Wilbur, running alongside, managed to coax the first powered airplane a few feet into the air above the dunes at Kitty Hawk. It would be nearly a half century later before a Russian-born immigrant named Igor Sikorsky designed and flew the first viable helicopter, thereby introducing the world to the wonders of unrestricted aviation.

A Brave New World

This is why the name Igor Sikorsky is universally known. Just think of all the neat things the *unrestricted* aviator can do with his remarkable freedom to slow down and hover. Because he no longer requires a runway to take off and land, he can operate his helicopter almost anywhere. All he needs is a clearing large enough to accommodate his aircraft, and he can come and go as he pleases. The possibilities are limitless.

Obviously his aircraft is a valuable military asset, but it can be used for humanitarian purposes as well. Because he can land anywhere, the helicopter pilot can use his machine as an airborne ambulance, dramatically cutting the time it would normally take to transport a patient by ground. If there's a car wreck, he can fly directly to the scene and land on the highway, right next to the accident. Then, while a team of medical experts cares for the injured accident victims in the back of his helicopter, he can fly directly to a trauma center.

Mount a hoist on it, and the pilot can employ his helicopter to rescue people trapped in places that would otherwise be inaccessible. Hang a bucket on the cargo hook, and his helicopter becomes an extremely valuable firefighting apparatus, especially over rugged terrain. Add infrared or night-vision equipment, and the helicopter can be used to search for people who are lost in the night, or even to find the bad guy who just robbed a convenience store. In the hands of a highly skilled pilot, there's virtually nothing this amazing machine can't do!

Ah, but there's a catch. Unfortunately, there's a price to be paid for this newfound freedom, and when things don't go as planned, the price can be very costly. As you're about to learn, this marvelous ability to slow down to a hover and perform all of these "neat tricks" is not without risk. This risk, however, has nothing to do with the common misconception that a helicopter's rotor, if it loses power, will stop turning and cause the helicopter to crash.

While it *is* true that a helicopter will take on the flight characteristics of a refrigerator if that big fan mounted on top of it stops spinning, the good news is this: Thanks to something called *autorotation*, the fan doesn't have to stop spinning just because the engine driving it has failed. Simply put, if the helicopter is moving forward at the "proper" airspeed, the wind moving across the blades will cause the rotor to continue rotating, just as an outdoor ceiling fan will turn without electricity as long as there's enough wind blowing across the porch where it's mounted.

Obviously, he can't continue to maintain his altitude after a total loss of power, but if the helicopter pilot immediately lowers the collective to decrease blade pitch and preserve the rotational momentum in his rotor, and if he adjusts his cyclic to maintain this *proper airspeed*—usually around 70 knots (a little over 80 miles per hour)—he can moderate the aircraft's descent rate, even with no power from the engine. As the helicopter descends, the relative wind across the rotor will autorotate the blades. It's true the helicopter is coming down sooner than he would prefer, but the pilot still possesses the ability to determine, within reason, where and how hard he will touch down. In a properly executed autorotation, the pilot pulls back on the cyclic as he descends below 100 feet or so, raising the nose of the helicopter. If he does this just right, the aircraft can be slowed to almost no forward speed prior to touching down—but this is where it gets dicey. In order to cushion the landing, the pilot then has to pull up on the collective (which, unfortunately, starts slowing his rotor), creating just enough lift to arrest the helicopter's descent rate—hopefully, before the rotor slows to the point that it can no longer produce the lift required to land safely.

The catch here is that the pilot *must not* raise the collective too soon or he'll bleed off all the rotational energy from the blades while he's still too high. When this happens, the blades stop turning too soon, and the aircraft will land hard, possibly hard enough to kill everyone on board. Conversely, if he pulls up on the collective too late, he won't be able to arrest the descent rate in time, and once again, he'll hit the ground hard. He *must* pull at precisely the right instant so that the rotor blades slow to the point where they're no longer generating lift just as the helicopter gently touches down. It's the same sort of timing you would use to spend the last dollar out of your bank account as you take your final breath.

Considering the way I've just described the steps involved, it may sound difficult, maybe even next to impossible, to execute a successful autorotational landing. This is certainly not the case. With enough practice, autorotations can become second nature to a skilled aviator. When he performs all of these steps correctly, the competent helicopter pilot can land his disabled aircraft on a dime.

Contrast this to the fixed-wing pilot who experiences a complete loss of power. He needs a clearing of sufficient length on which to land his gliding airplane and slow it to a stop—hopefully, before he crashes it into trees, buildings, or other obstacles in his path. Properly executed, the autorotation is a beautiful piece of airmanship, and because it requires less real estate, it's actually safer than trying to make a forced landing in an airplane.

"So, if that's true," you ask, "what's this increased risk that comes with performing 'neat tricks' in a helicopter? If a forced landing is safer in a helicopter than in an airplane, then what's the penalty the helicopter pilot must pay for being able to fly low and slow?"

Welcome to Helicopters 101

To answer the question of why a helicopter pilot is justifiably paranoid about losing an engine when he's flying low and slow, I need to solicit your patience while I digress even further into the world of academia. The digression is necessary because the key to your understanding the answer to this all-important question requires that you spend a few minutes examining a rather mundane-looking diagram on the following page. It's from a Bell-204 *Huey* flight manual, and the reason flying a helicopter, particularly a *public-safety* helicopter, can often be a risky proposition is because of the deadly consequence that can result from operating in the shaded area, to the left of the curved line.

LIFE INSIDE THE DEAD MAN'S CURVE

So, at the risk of making this chapter read more like a textbook than it already does, here we go:

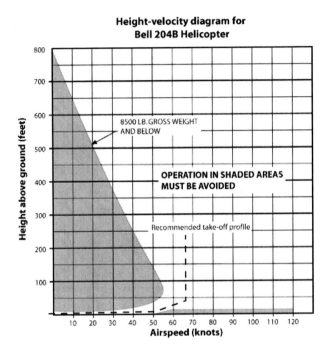

This is a typical example of something called a "height/velocity diagram," and it's truly a matter of life and death to helicopter pilots. As you can see, the vertical axis represents the height of the aircraft above the ground. This is plotted against airspeed, represented by the horizontal axis.

I already told you there's a "proper" airspeed (around 70 knots) at which to autorotate. This is actually the speed at which the rotor is most efficient. Think of it as the airspeed "sweet spot." Autorotate at a speed slower or faster than that sweet spot following an engine failure, and the rotor blades begin to slow down, decreasing the number of revolutions per minute (RPM).

Helicopter pilots call this droop in rotor RPM "bleeding turns," and it occurs because it becomes increasingly difficult to keep the rotor turning as the airspeed varies further from the sweet spot. Even during powered flight, with a normally operating engine, flying at anything slower or faster than that sweet-spot airspeed requires more power—power that may not be available to a public-safety pilot who happens to be flying over a fire with the additional weight of a 2,000-pound water bucket suspended below his helicopter.

The helicopter actually requires the greatest amount of power at two different airspeeds, zero and whatever the helicopter's maximum airspeed happens to be (usually around 150 knots). As the pilot accelerates out of a hover (zero airspeed), *less* and *less* power is required until he approaches 70 knots, at which point it now requires *more* and *more* power to fly faster. When there is no more power available from the engine, the helicopter has reached its maximum forward airspeed.

The more the airspeed differs from 70 knots, the more power it takes to keep the rotor turning at full RPM. In other words, and this is the critical point, it takes less power to fly at 70 knots than at any other airspeed because that's the airspeed at which the autorotational force produced by the relative wind is most efficient.

Now back to the chart. Keep in mind that the pilot of a helicopter that has suffered an engine failure, because he can't increase his power, can only accelerate by lowering his nose and diving toward the ground. Pilots call this trading altitude for airspeed. When a single-engine helicopter loses power, as long as the aircraft's altitude and airspeed come together in the white area of the height/velocity diagram, the pilot should have enough altitude to trade in exchange for the 70 knots of airspeed he needs to autorotate. This gives him a good chance to perform a successful autorotational landing.

In the case of a multi-engine helicopter, the white area represents the combinations of altitude and airspeed that allow the pilot to reach a speed at which the power from the remaining engine is sufficient to keep the rotor turning and allow the aircraft to remain airborne. Pilots of multi-engine helicopters refer to this as "flyaway speed." Flyaway speed is important because most multi-engine helicopters can't produce enough power to hover on one engine. Remember, hovering requires much more power than flying at 70 knots. This relatively safe one-engine-out flyaway speed usually occurs around 40 knots.

A Small Matter of Life and Death

So, what happens if an engine fails while the helicopter's altitude and airspeed meet in the dark-shaded portion of the chart? What if, for example, the pilot is flying his aircraft at 100 feet and 10 knots of airspeed? What if he is at 50 feet while hovering? Why is this height/velocity diagram such a big deal to helicopter pilots?

Well, the short answer is this: Whether you're flying a single-engine helicopter or a multi-engine helicopter, if you lose an engine—or even if you experience a situation where you suddenly need more power than the engine(s) can provide—while flying in the shaded area of the height/velocity diagram, you're probably going to crash. That's why the bold warning on the chart reads "OPERATION IN SHADED AREAS MUST BE AVOIDED."

And therein lies the rub. All of these wonderful things that a helicopter can do while flying low and slow (i.e., rescues, filling water buckets to fight fires, and landing in confined areas to pick up sick and injured people)—they all occur in the shaded area of that damned chart! That's why helicopter pilots refer to the curved line that separates the white and shaded areas of the height/velocity diagram as the *dead man's curve*. This is why you might want to excuse helicopter pilots for being just a little bit different.

In truth, the prudent helicopter pilot is constantly paranoid. And of all the helicopter pilots out there, public-safety pilots are the *most* paranoid. Why? Because, unfortunately, the area inside that *dead man's curve* is where they make their living. Every time a public-safety pilot comes to a high hover so the crew chief can send a rescuer down the hoist, every time he hovers 20 feet over a lake to fill his fire bucket with 2,000 pounds of water, every time he slows while descending 200 feet into a tight landing zone, he's on the wrong side of that thin line that separates the safe area of the chart from the "must-avoid" zone.

According to the Federal Aviation Administration's aircraft-mishap statistics, it takes fewer flight hours to kill a pilot flying public-safety missions in a helicopter than one performing any other operation in any other aircraft. Not since the U.S. Eighth Air Force was flying bombing raids over Europe has there been a costlier use of air assets in terms of the fatal mishap rate.

It's ironic, but flying a public-safety helicopter during rescue, firefighting, or emergency medical operations is one of the most dangerous, yet one of the most rewarding and exhilarating missions a pilot can ever experience.

It's *life inside the dead man's curve.*

4

Too Young for the Job

(Part One)

During my last two years in the Navy, I applied a full-court press to the person in charge of hiring pilots at Travis County. At least, at the time, I thought he was in charge of hiring pilots. Pat Leone, a former Army pilot, was the director of aviation operations for Travis County STAR Flight. I was determined to make sure he knew who I was, what my qualifications were, and just exactly how much I wanted to fly for STAR Flight. Every time I added another FAA rating to my résumé, I made sure Pat Leone was the first person to know about it. Every trip I made to Austin, I went out of my way to drop by and visit him in his office. I made sure that he understood I didn't just want to be an EMS pilot—I wanted to be an EMS pilot for Travis County.

I also spent those two years convincing Nancy that we wanted to raise our kids in Austin. We both had family in North Texas, near Fort Worth, and after ten years of living wherever the Navy happened to send us, she couldn't understand why I wanted to live in Austin instead of going back to the area she thought of as home. The truth of the matter is, when I was a midshipman at the University of Texas, I had already begun to think of *Austin* as home. The Central Texas hill country is one of the prettiest parts of the state, and when I left the Navy in 1992, Austin still had that college-town feel that made it seem more livable than other cities of comparable size.

Of course, the main reason I wanted to live in Austin was because I wanted to fly for STAR Flight. There were air ambulance helicopters based in the Dallas/Fort Worth area, but they were strictly EMS outfits. STAR Flight was a true public-safety program, not just an EMS helicopter. They flew the types of missions that were enticing to me for many of the same reasons that I had chosen to fly for the Navy. Call it a kid-like sense of adventure, I guess. At any rate, I had made up my mind that flying for STAR Flight was what I wanted to do, and I was pursuing it with the same sort of stubborn resolve that had served me well back in flight school.

As it turned out, it really wasn't all that hard to sell Nancy on the idea of living in Austin. After I took her there on a few weekend trips, she decided she liked the area for the same reasons I did. We even picked out a lot near Lake Travis, about twenty miles west of downtown. It had been repossessed from the original owner during the savings and loan crisis that took place during the late 1980s. The lot belonged to the Resolution Trust Corporation, a government-owned asset management company that had been set up for the purpose of liquidating foreclosure properties. Nancy and I were able to buy the lot for pennies on the dollar, and we started building a house on it almost a full year before my service obligation to the Navy was slated to end.

Looking back on it, I suppose it might have been a pretty bold decision to buy property in Austin while I was still in the Navy. After all, we had absolutely no assurance I would be able to land a job once we moved there. On the other hand, we knew that if we waited until I left the Navy, there was no way we were going to be able to qualify for a mortgage with no visible means of support, so we made the decision to dive head-long into the water and worry about whether or not we could actually swim later.

In the meantime, I continued lobbying Pat Leone, hoping he would be impressed enough to put me on his short list of candidates when it came time to hire another pilot. What I didn't know at the time, however, was that Pat Leone was struggling to hang on to his own job. For reasons unbeknownst to me, most of the pilots who flew for STAR Flight back then had lost confidence in his ability to lead, which was making it difficult for him to carry out his duties as the director of operations. Just about three months after I had separated from the Navy and moved to Austin, his chief pilot, who later told me he had grown tired of functioning as a heat shield between Pat and the line pilots, resigned and took a job in Alaska.

Pat Leone elected to fill the vacant managerial position from within the organization, probably hoping to establish a better rapport with the line pilots. Mike Phillips, who had joined STAR Flight shortly after the program's inception, became the new chief pilot, which meant there was now a vacant line pilot position up for grabs. Mike was a long-time Austinite, and like Pat Leone, he had been a helicopter pilot for the U.S. Army.

In fact, every pilot who had ever flown for STAR Flight was a former Army pilot. Mike had flown in Vietnam and was still flying AH-1 *Cobras* for the Army National Guard, where he was widely respected for his knowledge and expertise. Soft-spoken with a good sense of humor, Mike was well liked by just about everyone at STAR Flight, which made him an excellent choice for the chief pilot job. If anybody could mend the fences between Pat Leone and the other pilots, Mike Phillips would be the most likely candidate. What really mattered to me, though, was the position Mike had vacated to accept the job as chief pilot.

I had been watching the Travis County job listings like a hawk, and late in February, just a few weeks after I had come off active duty and finished moving to Austin, I spotted the advertisement for a STAR Flight pilot—right next to the listing for a heavy equipment operator. Within hours of the job having been posted on the county's television station (there were no websites in those days), I was on the phone to Pat Leone, who almost seemed surprised, if not slightly irritated, to hear from me.

This was more than just a little disheartening. After all, I had virtually taken up residence in his office since moving to Austin, and I was hoping for a much more enthusiastic reception now that he was actually in a position to offer me a job. The more I talked to Pat, the more I got the feeling I might not be at the top of his list of candidates to fill the vacant pilot slot.

Moreover, even if I had been at the top of his list, it turned out that the choice was not entirely his to make. The hiring process he outlined to me was not at all what I had expected. Mike Phillips had come up with a qualification matrix for selecting the top six candidates, who would be interviewed by an interagency committee of various STAR Flight personnel. The committee would comprise representatives from the three organizations that provided employees to the STAR Flight program. This meant that the interviews would be conducted by two paramedics from the City of Austin EMS Department, two nurses from Brackenridge Hospital, and two yet-to-be-determined representatives from Travis County. The committee would then select three candidates to fly with the STAR Flight instructor pilot, who would forward his recommendation, along with that of the committee, to Mike Phillips and Pat Leone, at which time the two of them would make the final decision on who to hire.

Of course, none of these wheels would even begin turning until the job had been publicly posted for two weeks, which only added to my frustration. The last thing Pat Leone told me before ending our phone conversation was that I "should direct all further communications with regard to the vacant pilot position to Mike Phillips." In other words, he was telling me to go away and quit bothering him.

To say I was disappointed after my phone call to Pat Leone would be a huge understatement. Of course, I'm not sure why I was so surprised. There was a lot of competition for good flying jobs—especially since the First Gulf War had just ended, and the military was dumping a ton of highly qualified pilots onto the civilian job market. I'm not sure whether I had been naïve, or just a little too cocky, but I had somehow allowed myself to believe that when a pilot slot eventually opened at STAR Flight, I would be the obvious candidate to fill it.

Now, confronted with a totally different reality, I looked back over the last couple of years and tried to figure out how I had mistakenly concluded it wouldn't be difficult to land the job I wanted. By the time I left the Navy, I had accumulated over five thousand flying hours. During my last two years on active duty, I had jumped through all the required hoops to earn an Airline

Transport Pilot certificate, the highest FAA rating a commercial pilot can hold. In fact, I held an ATP rating, not just for helicopters, but for multi-engine airplanes as well. In addition, I was a Certified Flight Instructor, a Certified Instrument Instructor, and a Certified Ground Instructor to boot. I had even received a Special Instrument Rating from the Navy, which meant that I could fly in weather that would ground most other pilots. I had punched every ticket I could punch to make my aviation credentials attractive to a potential employer. Throw in my irresistible charm, and I should have been a shoo-in.

Man, I felt like I needed to add "certified sucker" to my résumé. I should have known this wasn't going to be easy. Doggone it! I had allowed myself to believe I was the only guy qualified for my dream job, and now I was realizing that even though I might have been a big fish, I was swimming in a very large pond. And the pond was populated with a lot of other big fish as well. I simply hadn't counted on the fact that this was going to be such a competitive process.

Nancy told me I should try to remain positive, but I knew this wasn't good. In the first place, there just weren't that many EMS-pilot jobs around, and the one that had just become available at Travis County was one of the most desirable in the EMS universe. When it came to pay, benefits, and stability, STAR Flight was hard to beat. Add to that the fact that Austin was a good place to live, and it meant that during the mandatory two-week posting period, the folks at Travis County were going to be inundated with résumés from every helicopter pilot looking for a place to ply his trade—including those who were already employed by other EMS programs, a prospect I hadn't even considered to this point. This whole thing had been a huge miscalculation on my part.

TOO YOUNG FOR THE JOB (Part One)

Well, Pat Leone might have been tired of me already, but Mike Phillips was still a pristine target to this point. Up until now, I hadn't even shown up as a blip on his radar. That was about to change.

In the days that followed, you might say I made a regular nuisance of myself. Looking back on it now, I believe Mike Phillips deserved a medal for the tolerance he displayed in the face of my constant badgering. I had spent most of the last two years trying to convince Pat Leone that I would make a good pilot at STAR Flight, and now Mike was getting the two-week crash course on why yours truly was the best man for the job. I had never mastered the art of the soft sell, and why Mike didn't lose his patience with me after countless phone calls and numerous face-to-face visits, I'll never know. Most people would have been put off by the overbearing sales pitch I threw at him, but Mike took it all in stride and politely indulged my unadulterated hubris. He even took the time to help me understand the matrix he had designed for the purpose of ranking all the applicants.

Unfortunately for me, the matrix was weighted heavily toward someone who already had EMS credentials. No points were given for having had a résumé on file prior to the beginning of the hiring process, and even worse from my standpoint, it looked as if all of my additional FAA ratings weren't going to be enough to overcome my aggregate lack of EMS experience. And if all that wasn't bad enough, Mike let me know up front that my youth was going to be an issue. At thirty-five years of age, I was fifteen years younger than the average EMS pilot, and several crew members had already expressed some concern about flying with someone so young.

Sure enough, when the posting period was up, and Mike Phillips called to let me know that I hadn't made the list of pilots to be interviewed, my worst fears were realized. There was no way I could hide my disappointment, but I thanked Mike for calling to let me know what was happening. There were over a hundred applicants for the job, and I knew he wasn't taking the time to call every pilot who hadn't been selected for an interview.

Mike thanked me, in return, for my interest in STAR Flight and told me he would keep my résumé on file for any future openings. I could tell he was sincere, and I could also tell it was difficult for him to give me the bad news. He knew how desperately I wanted to fly for STAR Flight, and Mike Phillips was not the sort of person who enjoyed disappointing people. It was easy to understand why he was popular with the other STAR Flight employees.

Even though Mike Phillips had done his best to soften the blow, I took the news that I was out of the running for the open pilot slot extremely hard. Even though he had promised me I would get another chance the next time they needed to hire a pilot, I knew that it could be years before another spot opened up. Most EMS pilots are employed by companies, known as vendors, who supply the aviation assets to hospitals with an air ambulance program under a contract. If the contract doesn't get renewed when it expires, a new vendor comes in, and the pilots have to relocate to a different hospital, usually in a different city. That's assuming that the first vendor doesn't just let them go, in which case the pilots end up looking for another job.

Flying for STAR Flight was different. As county employees, pilots there enjoyed the kind of stability that airline pilots would envy. This meant the few slots available at STAR Flight were highly coveted positions to most helicopter pilots, and once a pilot was hired by Travis County, he rarely left of his own volition. This was a bad situation for Nancy and me. By building a house in Austin, we had put all our employment eggs into the STAR Flight basket, and now it looked as though I had missed out on the one opportunity I would have to land a job there.

Reluctantly, I began exploring other options. In contrast to the limited availability of EMS jobs at that time, there were numerous openings available for offshore helicopter pilots. Tasked with flying personnel and supplies to the oil rigs in the Gulf of Mexico, some of these pilots were based on the Texas Gulf Coast, but most of the time, flying offshore meant flying out of Louisiana. The work schedule was typically seven days on, followed by seven days off, so the prospect of commuting from our home in Austin wasn't necessarily a deal-breaker.

I also paid a visit to the lead pilot for CareFlite, the air ambulance operator in Fort Worth. It probably would have necessitated a move on our part, but if that was what it was going to take to land a job as an EMS pilot, we probably would have been willing to sell our house in Austin and relocate to that part of Texas.

As it turned out, however, we never got an opportunity to exercise that option. Even though CareFlite did have a pilot slot that had recently opened up, the people doing the hiring, just like the management at STAR Flight, were looking for someone who already had experience as an EMS pilot. I was beginning to wonder how it was ever possible to break in as a rookie EMS pilot, seeing as how every vacant EMS job could seemingly only be filled by pilots with prior EMS experience. Not only that, based on the demographics of the EMS pilots I had met, in addition to already possessing EMS experience, it appeared as if you had to be at least fifty years of age in order to become one.

Was there some sort of closed fraternity, in which flying jobs were a commodity that could only be traded with other members of the EMS pilot exchange? It certainly looked that way to me. I was only thirty-five years old and had no EMS experience, so it seemed as though I was ineligible to join the club. It was extremely frustrating, and it was also becoming apparent that I had been naively optimistic when I'd presumed my qualifications would earn me an EMS flying job right out of the Navy. My overconfidence notwithstanding, I knew for a fact that I had logged more flight time and possessed more ratings than most of the pilots who were already holding jobs with air ambulance operators—yet, because of my youth, I couldn't even get my foot in the door.

Then, just about the time I was reluctantly ready to pack my bags for the Gulf Coast, I received another phone call from Mike Phillips. It was immediately obvious from the tenor in his voice that, unlike the last time he had called me, he was about to deliver some good news.

And indeed he did. I listened as Mike explained that the committee had interviewed six pilots, but of those six, there was really only one with whom they had been impressed. When the committee's selectee went on to perform well during the in-flight portion of the interview, it looked as if the fact that they weren't interested in any of the other five pilots wasn't going to matter much. Mike Phillips and Pat Leone both agreed he was a good fit for STAR Flight, and they offered him the job—which he accepted.

That would have been the end of the story, but this is where it gets interesting. The pilot STAR Flight hired was the same pilot who had told his bosses at CareFlite that he was leaving to take another job. This explained the job opening in Fort Worth and further validated my closed-fraternity theory.

According to Mike, it had been a done deal until the newly hired pilot's family decided they wanted to stay in Fort Worth instead of moving to Austin. Because the committee was not impressed with any of the other pilots whom they'd already interviewed, everyone decided the best thing to do was to move on to the next group of pilots on the list and start the entire process again. This time, however, the committee would only be interviewing three pilots. If any of the three made the grade during the oral interview, only then would he actually be given a chance to fly.

Mike was calling to let me know that I was one of the three pilots who had been selected for the second round of interviews. This was on a Monday, and the interviews were scheduled for Thursday. I'm not sure why I decided (perhaps, because I was excited and nervous) that it was appropriate to offer him a flippant response to such an important question; but when he asked if I would be available, I told him that, "Even though I would love a chance to interview for the job, I already had plans to rearrange my sock drawer that day."

Fortunately, Mike laughed at my cavalier remark, but then he cautioned me, pointing out that I still had a big hill to climb because of my age. He told me I was at least ten years younger than each of the other candidates, and he also warned me that they both had a lot of EMS flight experience, which, not surprisingly, was going to be a major consideration to the interview committee. I told him I appreciated his candor and that I was happy just to have the opportunity to make my case in front of the people who would actually be making the selections.

Talk about getting a last-minute reprieve. I had all but given up on flying for STAR Flight in the foreseeable future, and now, out of the blue, it looked as though it could be a real possibility. I still had hurdles to clear, but the hurdles looked to be a lot less formidable than the brick wall that had been blocking my path just a few weeks earlier.

I couldn't understand why the pilot from CareFlite had opted not to take the job. Even though Mike said it was because his family had decided they wanted to stay in Fort Worth, I couldn't help thinking that, from a purely professional standpoint, he had made a bad choice. CareFlite was more established than most air ambulance operators, which is why I had tried to hire on with them when I was turned away at STAR Flight. Still, Travis County was the place to be for a helicopter pilot who was looking for a stable, long-term career in the EMS world.

I might not have been able to understand why my unwitting benefactor was staying at CareFlite, but I was absolutely ecstatic that he wasn't coming to STAR Flight. I couldn't help thinking that, because of the timing involved after my visit to CareFlite, he and I had probably passed one another somewhere along the interstate between Fort Worth and Austin, each of us heading in opposite directions between the same two jobs. Of course, there was still the small matter of the upcoming interview. Hopefully there would also be a check ride in my future—and after that, if all went well, maybe even an actual paycheck.

TOO YOUNG FOR THE JOB (Part One)

For Fans of the Iconic 1980 Movie The Blues Brothers

Just like Elwood, when he and Jake were putting the band back together, I was "on a mission from God" during the three days leading up to my interview. I started at the *Austin American Statesman*, the local Austin newspaper, where I dug up every newspaper article they had ever published on the STAR Flight program. There were a ton of them, too. STAR Flight had been around for seven years, and it was an extremely high-profile organization. Compared to today, there weren't nearly as many air ambulance programs around in 1992, and Austin was one of the smallest cities fortunate enough to have one of its own.

The flight crews were treated like celebrities by the local media, and I understood enough about celebrities to know they liked it when people reminded them of their accomplishments. I set about making copies of the articles that were most flattering toward the STAR Flight crew members and compiled them into a notebook. I also managed to acquire the names of the committee members from Mike Phillips. I figured it couldn't hurt if I just happened to show up with news stories written about the people who were actually interviewing me.

After that, I headed to the cleaners to drop off my best business suit, which was actually my *only* business suit. I had been cutting my own hair dating back to my days aboard ship in the Navy, but I went ahead and treated myself to a professional job at the local barber shop.

Lastly, and perhaps most importantly, I dusted off my old TH-57 NATOPS manual. NATOPS stands for Naval Air

Training and Operating Procedures Standardization, which is Navy-speak for the aircraft operator's manual, much like the owner's manual in the glove box of your car—except this one is the size of a New York City phone book. Assuming I made it past the oral interview, I would be flying a Bell 206L *Long Ranger* (STAR Flight's backup helicopter) on my audition with the instructor pilot. With only a few modifications, the TH-57, which was the helicopter I'd trained in as a student naval aviator, was a Navy variant of the Bell 206. This might have been an advantage for me if it weren't for the fact that just about every commercial helicopter pilot on the planet has flown some version of the 206 at some point in his career. There would be *one* thing, however, sure to work in my favor. In addition to training in it as a student, I had done a tour as a Navy flight instructor in the TH-57. I had spent several years teaching other pilots how to fly it, so if nothing else, I would be auditioning in a helicopter with which I was exceedingly familiar.

I didn't get much sleep the night before the interviews. I felt a little like an athlete trying to sleep before a big game. Trying to anticipate the questions I might be asked, I tossed and turned, formulating what I thought would be the perfect response to each query. Just about the time I had settled on the best way to reply to a potential question, I found myself wavering in favor of a better answer.

This turned into a seemingly interminable process during which I tried to cycle through all the possible answers to every possible question. Then I began worrying about what would happen if I actually made it to the flying portion of the interview. I hadn't flown in nearly six months, and even though I had logged well over five thousand hours, flying is a perishable skill. Then I worried about the Bell 206. I had never flown one, and although it was the same airframe as the TH-57, there were still a few differences in the two cockpits, albeit subtle ones.

This went on for most of the night. It was one trepidation after another. Mine was to be the third and final interview, so at least I didn't have to show up until after lunch, which turned out to be a good thing because I didn't manage to fall asleep until around four o'clock in the morning.

The interviews took place in a conference room at Signature Flight Support, which was located right next to the STAR Flight hangar at Robert Mueller Municipal Airport, just north of downtown Austin. Fixed-base operators, or FBOs, are like airport gas stations, only a lot bigger and nicer. In addition to renting out hangar space to local aircraft owners, they provide turnaround services to transient pilots, including pilots of military aircraft if the FBO happens to have a government fuels contract.

I had refueled at Signature many times over the years, so I was comfortably familiar with the surroundings. However, when I arrived at the gated entrance to the FBO parking lot, the Bell 206 was nowhere in sight. Instead, I saw the Bell 412, the primary STAR Flight helicopter, parked on the ramp out front.

This is not good, I thought to myself.

The 206 was the bird I was expecting to fly if I made it past the initial interview. I was intimately familiar with the TH-57, which was close enough to the Bell 206 to make me feel confident about my ability to fly it. I had absolutely no experience in anything that resembled a Bell 412. A little bit of panic started to creep in as I considered the possibility that I might be forced to take the most important check ride of my life in an aircraft about which I knew practically nothing.

Much to my relief, this turned out to be a non-issue. When Mike Phillips greeted me inside the FBO, I asked him why the 412 was there instead of the 206. He explained that one of the candidates from the morning interviews was out flying the 206 with the instructor pilot. The 412 was there because the on-duty pilot was a member of the interview committee. He and the rest of the crew had flown it over from Brackenridge Hospital earlier that morning.

While I was relieved to know I would still be flying the Bell 206, assuming I made it that far, the fact that it was already being flown by another candidate meant that at least one of my competitors had made it to the flying phase of the process. Mike and I chatted briefly in the lobby until I heard the door to the conference room swing open. Then someone I didn't recognize peeked his head around the corner and said they were ready for me. Mike shook my hand and wished me luck as I picked up my briefcase and headed off for my interview.

Once inside the conference room, I shook hands and introduced myself to each of the people standing around an impressively long table, a table that looked better suited for a board of inquiry than for a job interview. I had met several STAR Flight crew members on my trips to Austin over the past two years, but none of them were on the committee. During the introductions, I let the members know that I was grateful for the interview, and then I took my place—alone—at the far end of the table.

Everyone except me was dressed in a flight suit, and I couldn't help thinking how much more comfortable a "zoom bag" would have been instead of the business suit and tie I was wearing. I didn't know if the room was too warm, or if I was just nervous, but I could feel perspiration rolling down my arms inside my heavily starched shirtsleeves.

I don't remember many of the specific questions I tackled over the next hour and a half, but at least a couple of them dealt with subjects that provided me an opportunity to pull out the collection of newspaper articles I had gathered for just this purpose. From the reactions of the interviewers, I could tell they were impressed with the research I had done in preparing the notebook, and I also got the feeling that none of the other candidates had exercised enough foresight to do it.

Just as Mike Phillips had warned me, a few people on the panel voiced concern over my age, but I thought I did a pretty good job convincing those who were skeptical of my youth that, when it comes to safety, the skill and expertise of the pilot matter more than his age. There's a longstanding military aviation axiom that goes like this: "Gravity has no respect for age or rank." With that in mind, I explained to the committee that my youth was not indicative of my experience, as evidenced by the surplus of hours in my logbook.

The time seemed to be passing quickly, and—all in all—I was feeling pretty good about the way I was handling most of the questions they were throwing at me. After about an hour and a half, the committee chairman announced he was concluding the interview, and then he dismissed me to the lobby while he and the rest of the members decided whether or not I rated a chance to prove myself in the cockpit.

There was nothing I could do now but wait outside and hope.

5

Too Young for the Job

(Part Two)

When I emerged from my interview, Mike Phillips was waiting for me in the lobby. He told me to stand by, and then he disappeared into the conference room behind me and shut the door. I was cautiously optimistic that I would get a chance to fly, but as the minutes passed, I began to grow a little apprehensive. I knew a long jury deliberation was usually a good sign for a defendant on trial, but I couldn't understand why it was taking them so long to reach a decision.

Maybe I hadn't interviewed as well as I thought I had. Maybe some of the committee members just couldn't get past my youth and lack of EMS experience. One thing was certain. I had done everything in my power to prepare for this interview, and if it turned out I wasn't going to be granted an opportunity to fly, at least I had been given a chance to make my appeal directly to the people who were deciding my fate. After what seemed an eternity, Mike Phillips reappeared from the conference room.

I stood up in front of the couch where I'd been sitting, and Mike walked over to me, shook my hand, then asked me—"Are you ready to go flying?"

I'm sure he must have seen the relief on my face, and he chuckled a little as he asked me if I needed a flight suit. I told him I had one in my car, and I excused myself to go get it. As I was changing clothes in one of the hospitality rooms inside the FBO, I could hear the Bell 206 returning from the earlier interview flight, and despite my excitement and anticipation, I tried hard to maintain an even keel. I had told Nancy that I would call to let her know if I made it past the oral interview, but there was no time for that now.

The 206 had been gone for more than two hours, so I knew I was in for a pretty thorough check ride. That was okay, though. The hiring process was finally moving into the cockpit, which was where I was most comfortable. Now my future would hinge on the outcome of a flying competition, and maybe it was classic naval aviator bravado on my part, but there was only one pilot out there to whom I was afraid I might actually lose such a contest. Fortunately for me, my old buddy, Harold Graebe, was still in the Navy for a few more months.

TOO YOUNG FOR THE JOB (Part Two)

On the subject of natural-born pilots:

> *The best, most skillful pilot is the one who has the most experience.*
> —Chuck Yeager (test pilot)

One last word on the often-misunderstood issue of the naval aviator and the surplus of confidence he typically displays, frequently to the exasperation of those around him: You should know that he's not this way because he thinks he's fortunate enough to have been born with an innate skill that lesser pilots don't possess. Even the cockiest of naval aviators understands his abilities are the result of the intense training he received in flight school coupled with the experience gained in his fleet squadron. As I walked onto the ramp that day, the self-assuredness I felt going into my interview flight was the product of that training, along with the practical experience I had gained through logging as many hours in the air as most pilots ten years my senior.

Another way to look at it is to understand that I had already made a substantially greater number of mistakes than other pilots my age. It may seem counterintuitive to think of this as an advantage, but it truly is. Every mistake you survive carries with it a lesson. There is no such thing as a pilot who doesn't make mistakes, and the fact that he makes mistakes is not what makes a pilot dangerous. It's the pilot who doesn't realize he's making mistakes who is the true menace. I had recognized my mistakes, and I had learned my lessons well. Now it was time to demonstrate some of what I'd learned to the last person standing between me and my chance to become a STAR Flight pilot.

Up to this point, the competition to win this job had been a beauty contest, all about résumés, demographics, and prior EMS experience. Well, that was all in the past now. Now it was about how well I could shake the sticks. Now I was in my wheelhouse.

If Chuck Yeager's theory that there's no such thing as a natural-born pilot is true, then Merlin "Spanky" Handley was an enigma. He was certainly one of the best "seat-of-the-pants" pilots I ever met. He might not have been the most articulate of flight instructors, and he certainly wasn't the most proficient instrument pilot in the business, but his stick and rudder skills were second to none. During training flights, his extraordinary talents allowed him to challenge pilots under his tutelage with emergency scenarios that most flight instructors, myself included, wouldn't dare introduce—for fear that if the student didn't immediately recognize the situation and initiate the proper recovery, there wouldn't be enough margin in which to intervene. Spanky Handley was good enough to work within that tight margin and still prevent the *simulated* emergency from becoming an *all-too-real* mishap.

Aside from a brief hiatus, during which he left to fly for Hermann Life Flight in Houston, he had been flying for Travis County since STAR Flight's genesis, seven years earlier. Now the designated instructor pilot for the program, Spanky Handley had earned a reputation for challenging even the most competent pilot to perform under pressure. And as I was soon to learn, his reputation was well deserved.

As Mike Phillips escorted me to the just-returned Bell 206, heat from the turbine engine was still visible as it rose from the exhaust stacks. Spanky Handley had just finished debriefing the previous pilot candidate, who turned, shook my hand, and introduced himself to me. I'm ashamed to admit it, but two seconds after I had met him, I didn't remember his name. I was completely focused on the task at hand. As the other candidate and Mike Phillips walked back toward the FBO, Spanky, whom I had previously met during one of my visits to STAR Flight, asked me if I was ready. I told him I was, and he told me to strap in.

Our crew chief for the flight was Mike Kane, one of the senior STAR Flight paramedics, whom I had also met previously. Mike was young, friendly, and possessed a self-deprecating sense of humor that made him easy to like. He was the prototypical "type A" personality, a trait common to pilots and paramedics alike. His job that day would be to ride in the back of the aircraft so he could clear our tail rotor and, as we descended into LZs (landing zones), give me verbal commands to help me avoid obstacles.

As we got ourselves situated in the cockpit, Spanky briefly described the planned profile for the flight and offered to start the engine for me. Perhaps just a little too confidently, I let him know that I was familiar with the 206, so he told me to go ahead and fire it up. Suddenly, I was stricken by the realization that one of those *subtle* differences between the Navy TH-57 and the Bell 206 was the start procedure.

The TH-57 was equipped with a fuel control unit that automatically scheduled fuel flow during start, whereas the 206 start procedure required the pilot to *manually* control the rate at which fuel flowed to the engine until after it had completely accelerated to idle RPM. This involved opening and closing the throttle by means of a twist grip, much like the one on a motorcycle, which was integrated into the collective. To keep the turbine temperature within limits during this manual start, the Bell

206 pilot had to monitor the TOT (turbine outlet temperature) on the cockpit gauge and adjust the fuel-flow rate accordingly.

As the RPM of any turbine engine increases during startup, more air passes through the combustion chamber, and the rate at which fuel enters the chamber has to increase as well—otherwise, there won't be enough fuel to sustain the fire, and the engine will flame out before reaching idle. A flameout during startup is not a huge deal, but a *hot start* most certainly is. If too much fuel is introduced into the combustion chamber, the fire burns too hot, exceeding the TOT limits. If this occurs, the aircraft is grounded until a visual inspection to assess possible damage to the turbine blades can be performed by a qualified mechanic.

Suddenly I was facing a critical decision—suffer the humiliation of asking Spanky to start the aircraft for me (after I had just told him I was familiar with the 206), or risk damaging the engine and failing the most important check ride of my life without ever leaving the ground. The concept of a manually modulated start was not foreign to me, but I had never actually performed one. The prudent thing at this point would have been to admit that I needed help with the start procedure—which is, of course, exactly what I did *not* do.

It was probably not the smoothest startup Spanky Handley had ever observed, but I managed to nurse the gas turbine in that Pratt & Whitney engine all the way to idle RPM without flaming it out and, more importantly, without over-temping it in the process.

As I rolled the throttle the rest of the way to the full-open position and completed the remaining takeoff checks, I thought to myself, *Getting this thing started was the hard part. Flying it should be easy—that is, unless six months out of the cockpit have taken an unexpected toll on my flying skills.*

It didn't take long for me to realize that my skills had survived the layoff just fine. This was basically the same helicopter on which I had cut my teeth in flight school back at Whiting Field, and once we lifted into a hover, it felt like an old friend who was glad to see me again. As I hover-taxied my way from the ramp to the hold short line, I noticed Spanky Handley gradually moving his hands farther from the controls until they were finally resting on his knees.

Once we had departed the airport, Spanky had me shoot an approach to the Brackenridge Hospital helipad, which was only a couple of minutes away. I had made hundreds of approaches to small, *moving* flight decks in the Navy, and the approach and landing at the much larger, stationary pad at the hospital wasn't much of a challenge. Once I had successfully completed that, he told me to fly west, toward Lake Travis. While we were en route, he started asking me questions about how I would handle different emergencies.

There were a couple of times during the quiz when Spanky, after listening to me explain how I would respond to his hypothetical emergency, surreptitiously asked me, "Are you *sure* that's what you'd do?" When I was a Navy flight instructor, I had often used this technique on my own students just to see exactly how committed they were to their answers. It was an extremely effective technique for assessing a pilot's confidence level, even more so when his answer happened to be correct. In each case, I held firm to my initial response until Spanky was finally convinced I wasn't going to waver, even under his intense cross-examination.

This went on for about ten minutes, until we were over Mansfield Dam, a mammoth, concrete hydro-electric facility, separating Lake Travis from Lake Austin. Lake Austin, which is really nothing more than a wide section of the Lower Colorado River, is located 230 feet below Lake Travis.

Spanky asked Mike Kane to pick a spot on the ground, along the Lake Austin shoreline, which was to be my unprepared landing zone audition. Even though I was pretty sure this had all been scripted ahead of time, Mike pretended to weigh his options before fiendishly picking an LZ smack-dab in the middle of the giant, high-tension power lines that were strung between the dam and the substation below. The multiple layers of power lines, running in a half dozen directions, anywhere from 30 to 150 feet above the ground, were a menacing labyrinth; and Spanky, after asking me if I thought I could get in there, waited to see what I would do.

"I can get in there," I said. "The question is, just how badly do you want me to?"

"Pretend that's the only place to land, and you have to get in there or abort the call and leave the patient," he replied.

At this point, I wasn't sure if Spanky was trying to lure me into making a bad decision, or if he really wanted to see if I could get us in and out of the maze of power lines.

I told Mike to "clear my tail," and even before he had responded, I was going through my landing checks, setting up for my approach. Spanky just sat there and watched, so I figured, *He must really want to see if I can do this.*

Power lines are one of a helicopter pilot's greatest fears. Many a career, not to mention many a life, has ended as a consequence of "snagging a wire." As I write this book, I haven't flown for three years, and I still have a recurring nightmare in which I'm trapped between two layers of power lines. The power lines always seem to be getting closer together as I try to maneuver through them, and the only reason I haven't crashed in one of these dreams is because I always manage to wake up just in time. When I was still flying, I liked to think these dreams were God's way of reminding me to remain diligent in the cockpit. Now that I'm no longer flying, I think it's just His way of reminding me how fortunate I am to have survived a lengthy flying career that included more than my share of close calls.

As I called "rolling final" and cleared Mike to open the cabin door, I slowed my airspeed to around 30 knots. I was well below the face of the dam and only about 50 feet above the first set of wires, but I was still 200 feet above the LZ.

"Hold your altitude and keep coming forward," Mike said as he raised his voice over the rumble of the wind across his microphone.

As I crept forward over the wires, gradually slowing my ground speed to near zero, the gap that was the beginning of my intended approach path (between the wires and down to our LZ) was no longer visible in front of me. It had sunk behind the instrument panel, and after a second or two, it reappeared in the chin bubble, still partially obstructed by my feet as they rested on the pedals. I could feel the tension in my legs as I caught myself trying to push the pedals through the plexiglass.

Relax, I told myself. *You've done this a thousand times.*

"You're cleared down," Mike said.

With that, I lowered the collective just enough to begin a slow, almost-vertical descent toward the LZ, which was still 150 feet below. Now completely surrounded by wires, we had little margin for error.

"Your tail's still clear," Mike said in a reassuring tone.

Passing through 50 feet, the lowest and last set of wires disappeared from the chin bubble. I reacquired them out my side window, and they finally became visible through the canopy, just above the instrument panel. The hardest part of the approach was over, but I still had to set it down, and as we got closer to the ground, the steepness of the terrain was becoming more apparent.

Although the Bell 206 had no published limitations for sloped landings, it was not unheard of for pilots to lose control of the aircraft when landing on slopes greater than ten degrees. As I was making the approach into what was about 20 knots of wind, the uphill slope of the LZ was on my left side. This was significant because the 206, like most helicopters, hovered left skid low. This meant that if I landed facing into the wind, which is usually preferable, the left skid was going to touch down even earlier, and I would run out of lateral cyclic authority that much sooner. In other words, there would be a greater risk of allowing the aircraft to roll over on the ground, resulting in a crash.

"Clear my tail to starboard," I said to Mike as I arrested my descent about 10 feet above the ground.

Mike hesitated at first, then said, "Your tail is clear right."

After rotating my tail across the downslope side of the LZ until it was directly into the wind, I once again began a slow descent toward the steep incline.

"Right skid is down," Mike reported.

After briefly stopping the descent, I gently lowered the left skid until Mike reported it had disappeared into the tall grass. I stopped momentarily, one more time, and then lowered the collective just enough to plant the left skid gently onto terra firma just as the pointer on the attitude indicator moved a couple of degrees past the ten-degree benchmark.

"That was pretty good," Spanky said smiling, "but what's up with the Navy jargon?"

"What do you mean?" I replied.

"*Starboard?*" he asked wryly.

"So, what's wrong with that?"

"We're not in the Navy! We use 'left' and 'right'," he chided me.

"The crew chief is facing *aft* when he clears the tail rotor. How does he know whether I want to swing the tail to *my* right or *his* right?"

"Aft!"? Spanky asked, snickering. "We use 'forward' and 'backward' here."

This was an argument that would resurface time and again throughout my STAR Flight career, and I could tell I wasn't going to win it here. Instead, I changed the subject by asking Spanky if the candidate who had flown earlier in the day had elected to turn his tail into the wind, as I had done, or had he landed left-skid-first.

Spanky laughed and said, "He decided it was too risky to come down through the wires, so he didn't even shoot the approach."

At first, I wasn't sure if the other guy had made the better decision in Spanky's eyes, but as soon as he grinned and gave me a thumbs-up, I knew I had done well in accepting the challenge.

"Now, can you get us back out of here?" he asked.

"What's next?" I replied.

"Let's take off and head west again."

The departure through the network of wires, though not as nerve-racking as the approach and landing had been, still demanded a steady hand. After coming straight up a few feet, I stopped my climb and made a slow pedal turn to get us back into the wind, this time swinging the tail *left* to avoid the higher terrain, which was now on my right side. Once I had finished the turn, I briefly looked down at the torque gauge and raised the collective to the max-power setting. Quickly shifting my scan back outside the cockpit, I began the ascent through the maze of high-tension power lines until I was just clear of the tallest set of wires, at which point I lowered the nose and began transitioning to forward flight.

Once Mike had called the tail rotor clear of all the power lines, I lowered the nose a little more and began a descent right over the middle of the narrow lake. As long as there are no obstacles in the way, trading altitude for airspeed is the best way for a pilot hovering at 200 feet to get out from behind the *dead man's curve.*

"I like that," Spanky said approvingly as we descended below the tops of the hills adjacent to the narrow lake. "Go for the airspeed first."

Once I had accelerated straight ahead to around 90 knots, I pulled back on the cyclic and climbed until we were clear of the surrounding terrain, at which point I began a slow turn to the west. This took us back across the dam and out over Lake Travis as we continued to climb. Spanky told me to level off at 3,000 feet and pointed to the highest part of a ridge, which looked to be about ten miles in front of us.

"That way," he said.

By the time I reached 3,000 feet on the altimeter, we were over the western shore of the lake. I lowered the nose, increased my airspeed to 120 knots, and steadied up on a course straight toward the peak on the ridge. We were about an hour into the flight at this point, and it was then that Spanky unexpectedly engaged me in some casual conversation. He started asking me questions about my personal history.

"Where did you grow up?"

"Where did you go to school?"

"What did you fly in the Navy?"

This went on for a few minutes until we were just a mile or two from the ridge. Looking at Spanky, I was right in the middle of answering one of his questions when, without taking his eyes off me, he grinned and suddenly rolled the throttle to the idle position.

He had just given me a simulated engine failure.

In every moment of choice, you create a new destiny.

—Kevin Michel (author)

As soon as I felt the nose yaw to the left from the abrupt power loss, I instinctively lowered the collective to maintain our rotor RPM and pulled back slightly on the cyclic. I began slowing to the minimum rate-of-descent airspeed, which was 60 knots. Our indicated altitude of 3,000 feet put us about 2,000 feet above the ground, so I didn't have much time in which to find a suitable clearing for the autorotation.

Not only that, Spanky had purposely initiated the emergency while I was focused inside the cockpit to ensure I wouldn't already have a good landing spot picked out. I didn't see anything promising in front of me, and since I was flying downwind when the simulated engine failure occurred, I began a slow turn to the right to at least get into the wind. This would facilitate a quick search of the ground behind me, where I hoped to find a clearing and set up for an approach.

Just as I initiated my turn, however, I noticed a short, abandoned airstrip at our two o'clock. This *had* to be the spot Spanky had intended for me to select when he'd initiated the emergency, so I rolled out of the turn, with the wind still on my tail, and headed for the overgrown, asphalt runway. From where I was, on my new heading, I was staring right down the faded runway centerline. I knew I had enough altitude to get there if I held my present heading and made a straight-in approach. The only problem was, a straight-in approach meant that I would be landing with a 20-knot tailwind.

Knowing I might not have enough altitude to set up for an approach into the wind, but needing to make a decision quickly, I lowered the nose and accelerated back to the maximum-range airspeed. I was hoping to extend my glide as much as I could toward the far end of the runway. I was doing a good job managing the rotor RPM, which minimized our descent rate, but I had to decide right then and there—continue the straight-in approach and land downwind, or set up for a turn and risk coming up short. I decided to press my luck. Turning away from the runway, my plan was to execute a one-hundred-eighty-degree turn back into the wind at the bottom of the autorotation.

No longer headed straight for the only spot where I could land my autorotating helicopter, I was offsetting my initial approach to the left, hoping to leave myself enough room for a last-second, right-hand turn back to the runway. The abandoned airstrip was the only clearing available to me, so if the autorotation was going to end successfully, I had to make it back to the runway coming out of the turn.

The engine failure was simulated, of course, so it's not as if this was a death-defying stunt. There was always the option to roll the throttle back up and terminate the autorotation. If, at any point during the approach, Spanky didn't like what he was seeing, he could tell me to wave it off, or he could take control and do it himself. This is one reason I had decided to raise the stakes by setting up for a turn, even though it meant I might run out of altitude before reaching the runway.

After all, the only thing riding on the outcome of my decision to offset the approach was the competition to win the job at STAR Flight. That said, winning the job at STAR Flight, even though it wasn't a matter of life and death, was still a pretty huge deal to me—especially considering how hard I had worked to get to this point. In the heat of the moment, I had made a high-risk, high-reward choice. I knew that if I botched this one maneuver, it might ruin my chances altogether, but if I pulled it off, it would probably seal the deal in my favor.

I continued heading out over the trees, hoping I would still have the altitude I needed to get the helicopter turned back into the wind with enough pavement available on which to complete the autorotation—assuming Spanky didn't tell me to roll the throttle back up beforehand. He still hadn't told me if I was flying the autorotation all the way to the ground or making an early power recovery.

In less than a minute from the time the event had started, I reached a position abeam the nearest end of the runway with just enough altitude to make the turn—but not enough to extend my glide any farther. Even at this late stage, I still could have aborted my plan by sliding right and settling for the downwind landing on the asphalt, but I knew it would take a few seconds to complete a course reversal. My instincts told me those few seconds would be enough for the 20 knots of wind on my tail to continue pushing the aircraft far enough down the runway to give me the room I needed.

Immediately, I banked the aircraft hard to the right. If I had indeed been right when I estimated the effects of the wind on our turn, we were about to roll out directly over the runway with just enough room to complete the autorotation. Halfway through the turn, Spanky must have liked what he was seeing, because he told me to take it all the way to the ground.

Fortunately, my instincts were validated as we rolled final with just enough altitude and runway. A split second before I was out of the turn and aligned with the runway centerline, I raised the nose, grabbed an armful of collective, and managed to coax the decelerating helicopter into a perfect, feather-soft, zero-groundspeed touchdown, just as we were about to run out of pavement and rotor RPM.

Relieved, and more than a little pleased with myself at having just consummated a textbook autorotation, I lowered the collective to the full-down position. The idling engine once again began to drive the rotor, which—by this time—was turning slowly enough for the three of us to see the individual blades as they rotated around the now-stationary helicopter.

Just then, Mike Kane keyed up his microphone from the back seat and emphatically announced, "You've got the job!"

I looked over at Spanky Handley to see what his reaction was going to be.

It had been almost two months since Mike Phillips had called to inform me that I'd been selected for an interview. Forty-eight hours after I had completed my check ride with Spanky Handley, Mike had called me again—to let me know that the STAR Flight job was mine. It had taken two days for Mike and Spanky to convince Pat Leone that he should take a chance on a young pilot with no EMS experience.

The weeks that followed had been filled with training flights and countless hours of study, all leading up to a check ride with an inspector from the Federal Aviation Administration. Compared to the one I had flown with Spanky, the check ride with the FAA inspector had been a cakewalk.

So, there I was. I was finally a STAR Flight pilot.

As a matter circumstance, my first scheduled shift after being cleared to fly the line happened to fall at the beginning of a four-night hitch. Back when I was hired, Travis County pilots worked "four-on and four-off" (four twelve-hour day shifts, followed by four days off, then four twelve-hour night shifts, followed by another four days off). It was early May, so at 7:00 p.m., when the shift began, there was still some daylight left. Daytime flights are usually less stressful, so I was hoping we would get dispatched before nightfall so I could get that first call under my belt before it got dark.

I might have been brand new to the job, but the crew members scheduled to fly with me on that first shift were a couple of seasoned veterans. Ron Startzel was a hard-drinking chain smoker, who looked ten years older than he really was. Ron had been a STAR Flight nurse for several years and didn't seem all that pleased to be flying with a rookie pilot. The other member of my crew that night just happened to be the chief flight medic at the time. Dave Williams was a young, easy-going guy from Fort Worth, and although our paths had never crossed prior to STAR Flight, the two of us had grown up not too far from one another. Dave's pleasant demeanor was a welcome contrast to Ron's surly disposition. During my training, Dave had gone out of his way on more than one occasion to help me and make me feel like I was part of the team.

Unlike the pilots, STAR Flight medical crews worked twenty-four-hour shifts back then, so Ron Startzel and Dave Williams had already been on duty for twelve hours by the time I reported to the hospital. Even though, as the oncoming pilot, it was solely my responsibility to preflight the aircraft at the beginning of my shift, Dave followed me up to the helipad and helped me to check out our bird. Of course, because this *was* my first shift, he might well have wanted to make sure I knew what I was doing, but he certainly didn't let on like that was the case.

By the time Dave Williams and I made it back down to the crew quarters, the daylight was fading, along with my chances to get a call under my belt before dark. Ron had already retired to the back bedroom, and as Dave and I sat outside on the patio, we chatted about what it was like to work for STAR Flight.

It was a perfect spring night. From our vantage point leaning back in our chairs, we had a spectacular view of the brightly illuminated dome on the Texas State Capitol, two blocks away. It was absolutely perfect, and I couldn't believe how lucky I was to be sitting there. Just a couple of months earlier, it looked as if I was going to have no chance at all to fly for STAR Flight, and now I was savoring the moment. There was no place else I wanted to be just then.

After a couple of hours, Dave decided to turn in, leaving me as the only crew member who was still awake. I was a little apprehensive about getting too comfortable, so I elected to stay up. I wanted to make sure I would be alert if we were called upon to fly in the middle of the night. After occupying myself with some map study for about an hour, I settled onto the couch for some all-night television, a routine I would repeat often over the next

TOO YOUNG FOR THE JOB (Part Two)

twenty years. Everything remained quiet until just around 3:45 a.m., when, right in the middle of my fourth episode of *Hogan's Heroes*, the pager went off. I was about to fly my first call as a STAR Flight pilot.

There is always some specific moment when we become aware that our youth is gone.

—Mignon McLaughlin (journalist)

I had ridden along on several dozen EMS calls while I was training, mostly during daylight. Even so, I had been exposed to a few nighttime calls, enough to know that once you got within a few miles of the scene, it was usually easy to spot the overhead emergency lights on the first-responder vehicles. As we flew through the pitch-black darkness of a moonless night, I was relying heavily on that fact to help me find the scene. We were headed to a motor vehicle accident (MVA) on a lightly traveled farm-to-market road, forty miles west of Austin.

According to our dispatcher, who was communicating with us on the radio, the caller who reported the accident had driven up on the scene of a head-on collision. Without even getting out of his car, he had turned around to find a telephone so he could call 911. This was 1992, and not everyone carried a cell phone back then. Even if someone was fortunate enough to have a cell phone in his possession, there often wasn't a tower within range, especially in an unpopulated, rural area.

Flying in the dark with nothing more than a road map (GPS was nonexistent), I was forced to fall back on an archaic form of navigation known as "dead reckoning." Essentially, it's a technique whereby the pilot flies a constant heading, and then based upon the elapsed time from his last known fix, he estimates his position by plotting a straight line on a navigational chart.

Well, my navigational chart had been purchased at a Texaco gas station, and my last known fix was Brackenridge Hospital, some forty miles and twenty minutes behind me. It was also dark—really dark. The roads and other landmarks we needed to accurately determine our position were nothing more than rumors in the night. We were literally flying blind over some of the roughest terrain and highest elevations in Central Texas.

To make matters worse, both the ground ambulance and the state trooper assigned to the call were reportedly still miles from the area where, based upon the information we had, we expected to find the accident. It was starting to look like we would be the first to arrive on the scene—assuming we could find it.

I asked Dave Williams, who was riding in the copilot seat, "How often do we beat the ground units to a call?"

"We don't," he replied. "This is a first for me."

"How do they expect us to find the scene?" I asked.

Dave looked at me and laughed, but Ron Startzel was not amused.

"Do we have a clue where we are?" he asked sarcastically from the back of the aircraft.

As much as I didn't appreciate the derisive tone with which it was delivered, Ron's question was a good one—and one for which I didn't have a good answer. Because we couldn't see the topography around us, we were flying 1,000 feet above the terrain, just to be safe. Unfortunately, without the emergency overhead lights from the ground units to guide us, we weren't going to find the accident site from that altitude, so I decided to slow down and descend, using the night sun to clear a path in front of the helicopter.

An industrial-strength searchlight mounted on the underside of the helicopter, the night sun was the best tool we had for conducting low-altitude searches at night. It generated thirty million candles of light and was remotely controlled with a four-way thumb switch mounted on the collective. The pilot could rotate it through one hundred and eighty degrees, point it ten degrees up or straight down, and vary the width of the beam without ever taking his left hand off the collective.

This was important because it allowed me to sweep the area in front of and below the aircraft as we descended to 200 feet above the ground. If we were lucky, we might stumble upon the road on which the accident was reported to have taken place, and then, assuming we chose the correct direction to fly from that point, we could follow it to the scene. From 200 feet, our maximum field of vision from the night sun was about the size of one city block. Finding the road was still going to be like finding a needle in a haystack, but at least we could see the haystack now.

A few minutes later, it appeared out of nowhere. We had not only found the road—we were directly on top of the accident site.

"Bingo!" Dave said.

"It's about time," Ron grumbled.

I'd like to be able to say that our finding the accident site was the result of my superior navigational skills, but in reality, it was pure, dumb luck. We just stumbled onto it.

The celebration was short-lived however. The relief I felt from having located the accident quickly faded as we surveyed the ghastly spectacle below. Even from 200 feet, we could tell there were likely no survivors. Eerily illuminated by the unsteady beam from our night sun, in an area otherwise void of ambient light, the two cars were twisted and crushed beyond anything I had ever seen. It was easy to understand why the unfortunate soul who'd happened upon this grisly scene had immediately turned around without stopping.

As we circled overhead, the ambulance crew reported over the radio that they were still fifteen minutes away. The state trooper was not any closer. During my training, I had been instructed not to land on roadways until they were secured by someone on the ground, so I continued circling for another minute or so, not sure what to do.

"Are we going to stay up here all night?" Ron asked out loud.

I looked for an LZ off to the side of the road, but there simply wasn't enough room because of the rugged terrain. However, both cars had come to rest in the middle of the narrow, two-lane road, and because they were separated by a hundred feet or so, I decided it would be safe to land between them.

The wrecked cars, I reasoned (although I had no real experience on which to base my conclusion), would essentially serve as barricades to prevent other vehicles from running into the helicopter while we were on the ground. My major concern was picking up crash debris in our rotor wash on short final. The vortices generated at the tips of the main rotor blades sometimes lift loose objects from the ground, then suck them back down through the rotor disc, which can damage the aircraft. The pilot can usually minimize this risk by executing a quick, no-hover landing, which is what I elected to do. Even though we blew some debris off the road, we managed to avoid sending it skyward. Still,

I expedited the shutdown, just to be on the safe side. Dave and Ron both exited the helicopter as I applied the rotor brake, switched on the scene lights, and opened the pilot-side door.

I was immediately aware of two things as I climbed out of the cockpit: First, aside from the sound of hissing steam that was coming from one of the wrecked cars, it was deathly silent. Second, there was a sickly sweet odor in the air from the still-hot engine coolant on the pavement. The scene lights, one floodlight on each side of the aircraft, revealed the horrific aftermath of an incredibly violent, high-speed collision.

As I walked toward Dave Williams, who was inspecting the car closest to where we had landed the helicopter, I could now see the steam I had previously only heard, which—strangely—was emanating from *inside* the driver-side door, or what was left of it anyway. Dave turned toward me and, with a somber expression on his face, reached for the microphone attached to his shoulder strap. He radioed the dispatcher that there were three victims in the vehicle, all DOS (dead on scene).

"You can tell just by looking?" I asked, stopping several feet short of the wreckage.

"Look for yourself," he said, and then he shined his flashlight inside the hideously distorted hunk of metal that had once been a car.

Without thinking, I stepped closer, looked inside the wrecked car, and then immediately wished I hadn't. There are no words I can use to describe what I felt when I saw the condition of the three corpses, just feet from where I was standing. The two who were grotesquely entrapped in the back seat of the wreckage appeared to be teenaged girls, but it was impossible to tell for sure. The adult in the front of the vehicle wasn't immediately recognizable as a human body, but I could tell from his clothing that he was a male. He was pinned between the steering column and the engine block—which was now in the front seat.

Just then, I realized that the steam coming from inside the car carried with it the sickening smell of burned human flesh. The victim in the front seat was literally being cooked by the hot engine block. Slightly nauseous, I turned away and leaned forward to brace my hands on my knees until the feeling had passed.

"Don't worry," Dave said to me. "You *won't* get used to it."

Dave probably didn't realize it, but his statement was immediately corroborated by the look I saw in his eyes. He was shaken as well—maybe not as visibly as I was, but shaken nevertheless.

By this time, we could hear the not-too-distant sound of the state trooper's siren, as well as the one from the ground ambulance, which was approaching from the opposite direction. Soon after that, as they pierced the darkness on either side of us, we could see both sets of overhead lights flashing from well beyond the glow of the helicopter's scene lights. I found myself becoming more and more unnerved by the stereophonic wailing of sirens, steadily increasing in volume as they converged on us from both sides of the hellish wreckage. The macabre surroundings served to make the experience seem even more surreal to me than it already had to that point.

I don't know why, but I found myself wishing someone else would hurry up and get there. Obviously, we weren't going to be transporting any patients. I already knew the people in the other car had to be dead too. There was absolutely nothing anyone else could do to help, but I wanted them to get there anyway. I remember thinking to myself, *The sooner they arrive, the sooner we can leave.* I had seen enough, and I was anxious to be somewhere else—anywhere but there.

I never went to the second vehicle, but I heard Ron call in two more DOS confirmations over the radio. He appeared to be largely unfazed by the mayhem as he matter-of-factly informed Dave and me that the victims in the other car were an elderly man and woman. Judging from the lack of skid marks at the point where the impact had obviously occurred, it looked to me as if the elderly couple, after drifting into the oncoming lane, had slammed into the other vehicle without ever applying the brakes. But then again, the whole scene was one giant debris field. I couldn't even be sure which car had been traveling in which direction. At any rate, the combined speed of the two cars had to have been well over 100 miles per hour at the moment of impact.

The ambulance finally showed up, and as Dave and Ron were turning the scene over to the two ground paramedics, I found myself wondering if they were the ones who would be left with the grim chore of untangling the bodies from the wreckage. Then the state trooper arrived, and I wondered if he would be tasked with notifying the victims' families.

How is he going to tell them about what just happened here? I thought.

There wasn't much conversation on the flight back to the hospital, although as soon as we had cleared the LZ, Ron Startzel interrupted the silence.

"You still want to be a STAR Flight pilot?" he asked me.

I sensed a slightly sadistic pleasure in his voice, almost as if he was taunting me because he knew how badly I had wanted this job and how hard I had worked to get it. I wanted to tell him what a miserable jerk he was, but then I decided I was in no mood to engage Ron in an argument, so instead, I simply chose to ignore the question.

Once again, however, even though I didn't appreciate the spirit in which Ron had asked it—his question was one for which I didn't have an immediate answer.

Perhaps it was because I wasn't yet accustomed to working night shifts. . . . Maybe it was because I had trouble forgetting the nightmarish images I had seen a few hours earlier. . . . Regardless of the reason, when I got home the next morning, I knew there was no use in trying to sleep.

After coming through the front door, I went straight to the laundry room. The smell of burned flesh, along with the other haunting scents from the wrecked cars, had permeated the fabric of my flight suit, and I couldn't wait to strip it off. The county-issued, fire-resistant garment was expensive and relatively new, so I knew I couldn't simply throw it away (which was what I would have preferred). Instead, I opened the lid to the washing machine, tossed it in, and immediately slammed the lid shut. I didn't even stick around to start the machine. I know that seems irrational, but it would have taken a few more seconds, and I just wanted to get away from the nauseating odors.

Unfortunately, even after leaving the laundry room, the aura from the sickening scene I had witnessed on that dark (now sinister to me) stretch of roadway still stubbornly clung to me. I desperately needed to sanitize my senses, so I quickly headed up the stairs, relishing the thought of a long, unabridged shower.

I had to go through the master bedroom, where Nancy was still sleeping, and although I tried not to wake her, she must have heard me come in.

"How did it go?" she asked me.

I quietly closed the bedroom door and hurried toward my anticipated liberation from the lingering stench.

"I'll tell you later," I said. "Go back to sleep."

I must have stood there under the shower for twenty minutes, but when the hot water ran out, and I turned it off, . . . the burned flesh, the hot engine coolant, and what I can only describe as the smell of death were still clearly palpable in my mind. It was as if the odors (and the impressions associated with them) were now indelibly etched into my psyche. I didn't know it yet, but this was a drill I would repeat far too often in the coming years. This sort of anxiety was something for which I hadn't bargained, and unfortunately for me and my now-distant youthful innocence, Dave Williams had been right. I never did get used to it.

As I stood there that morning, reflecting on my first shift as a STAR Flight pilot, there was one thing I knew for certain. Through the echoing silence of that steam-filled shower stall came the realization that if any of my new colleagues were still concerned that I might be too young for this job, they could rest easy.

By the time I reported for my second shift, less than twelve hours later—I had already aged well beyond my years.

6

Anatomy of a STAR Flight Pilot

As I write this book, it would no doubt be easier to explain my life as a STAR Flight pilot were I able to dissect it. The problem here is that I need to quadrisect it. Even though I can barely spell *quadrisect*, and despite the fact that I'm only moderately confident I know what it means, I'll give it my best shot. I guess the point I'm trying to make here is that, while it's true that a majority of the flights that were assigned to STAR Flight were garden-variety EMS missions, it would be an oversimplification to think of it strictly as an EMS operation.

In addition to the countless EMS calls to which we responded, we also flew rescue, firefighting, and law enforcement missions. The piloting skill-set we had to maintain in order to be proficient at our jobs comprised a fairly wide spectrum. The same is true for the medical crew members who flew with us. They were more than just medics and nurses who happened to be wearing flight suits.

STAR Flight crews had to be well versed in other public-safety genres as well. EMS calls certainly required a high level of skill from the onboard medical personnel, but from the pilot's perspective, these flights represented lesser challenges than did the other types of missions we flew. Fire suppression and rescue missions constituted much more hazardous flight regimes and were, therefore, much more exhilarating for the crew member who happened to be sitting in the pilot's seat. These were the missions that truly required the guy driving the helicopter to be at the top of his game. When fighting fires or conducting rescue operations, the pilot contributed greatly to the ultimate success or failure of the mission. It should come as no surprise that these were the types of calls I most enjoyed.

Conversely, even though EMS calls sometimes involved high drama, the pilot, once he had made the decision to launch, was rarely critical to the outcome. Of course, as the pilot in command, I was responsible for the safety of everyone aboard the helicopter, so the weight of responsibility was on my shoulders every time we left the pad; but as long as the weather was good and the helicopter didn't break, my intensity meter often fluctuated only moderately during a routine EMS flight.

As for the law enforcement calls, with a few notable exceptions, my intensity meter frequently registered a negative value. Had we been dispatched in a timely manner, this might not have been the case; but I can't begin to count the number of times we were called in after a pursuit had, for all intents and purposes, already ended. Only after the suspect had jumped from his car, eluded the pursuing officers, and escaped into the night on foot would we be called in to join the effort.

Too often, we would arrive on the scene and be given a last-known location from which to begin our search, only to learn that it had been anywhere from thirty minutes to an hour since the suspect had last been seen. Even with night-vision goggles, the odds of finding someone who had been moving for that long were extremely remote. Still, we did manage to collar our share of bad guys from time to time, but the number of successful searches were miniscule compared to the number that ended in futility.

Although they were few compared to the boring ones, most of the *successful* law enforcement missions were the ones to which we were dispatched while a pursuit was still in progress. If we could get there before the bad guy had managed to evade his pursuers, his odds of getting away diminished greatly. Once we had a visual on his vehicle, he couldn't outrun us, no matter how hard he tried. Often, the suspect would give up almost as soon as we arrived.

There was one night, however, when, instead of convincing him to surrender, our arrival on the scene of a high-speed chase might have precipitated a deadly confrontation between the fleeing suspect and his pursuer.

The real "Dirty Harry" is not a homicide inspector with the San Francisco Police Department. He's a deputy sheriff in a rural Central Texas county.

There were a few law enforcement officers out there who recognized the advantage they gained by calling us out early in a pursuit. One of those officers, a deputy from an adjacent county, attempted to make a traffic stop one night on a remote stretch of highway about twenty miles from Austin. It was just after two o'clock in the morning (right after the local bars had closed), and the man the deputy tried to stop was driving erratically in a pickup truck. When the deputy switched on his lights and siren, the man slowed down at first—then decided to hit the gas instead.

As soon as the deputy started his pursuit, he radioed for STAR Flight, and within ten minutes, we were close enough to see his overhead lights just a few miles in front of us. As soon as I turned on the night sun and trained it toward the fleeing pickup, the suspect, without even pulling over, slammed on his brakes, stopped his vehicle right in the middle of the road, and turned off his engine and headlights (remember, this is where they usually give up). The deputy stopped his cruiser, also in the middle of the road, about two hundred feet behind the truck.

Suddenly, we heard the deputy yelling "shots fired!" over the radio, and then everything went silent. We tried to reestablish communications several times, but the deputy wasn't answering. When we arrived overhead a few seconds later, we could see one person lying prone on the pavement and the other walking toward him, aiming a handgun. At this point, we didn't know who was who, so we tried again to reach the deputy over the radio—still no response. We were eventually able to determine that the person lying on the pavement was not wearing a uniform (in fact, he was shirtless), so we assumed the person who was still standing had to be the deputy. Even though we still had no communications, we knew it was likely that at least one of them had been wounded in the shootout, so we decided to go ahead and land.

It turned out that the man driving the pickup had decided it was time to end the chase. I can't be sure, but I suspect that when he saw our night sun approaching, he knew he wasn't going to get away—so instead, he stopped his truck, jumped out with a shotgun, and began firing as he moved brazenly toward the deputy.

The deputy stepped out from his car and took cover behind the driver-side door. He calmly waited for the shotgun-wielding bad guy to get to within thirty yards or so, then squeezed off a single round from his nine-millimeter service weapon. It struck the suspect dead-center between the eyes and dropped him right there in the middle of the highway.

After we landed, my medic and nurse attempted CPR until they eventually received a death pronouncement, via cell phone, from a physician. While my crew was working the patient, the deputy showed me where the pellets from the shotgun blasts had struck his vehicle several times as he was standing behind the door. I asked the deputy how many return shots he had fired during the exchange, and he answered that he had fired just once.

"This is the most impressive thing I've ever seen," I said. "You must have ice water in your veins to pull off a shot like that while you're under fire."

"Not really," he said matter-of-factly. "I was aiming center mass."

The Point of Diminishing Returns

Needless to say, most of our law enforcement operations were not nearly so intense. To the contrary, because we were routinely called to the chase too late, most of them were incredibly boring—a complete waste of time and fuel.

As for the EMS calls I ran, they were often compelling, but as I stated earlier, the competence of the medical crew was more critical to the patient's outcome than the skill of the pilot. However, when it came to fighting fires and flying rescue missions, it was all about flying the helicopter. The ultimate success or failure of these missions was a function of how well I could shake the sticks, and that's what made flying for STAR Flight much more rewarding than flying for a typical air ambulance program.

If I could simply talk about the stick-and-rudder stuff, apart from the human side of the experience, it would be much easier to paint an accurate picture of my career as a public-safety helicopter pilot. Here's the thing, though—the two are inexorably intertwined. It's impossible for me to analyze and explain them independently. The truth is, my passion for flying was not the only reason I looked forward to my shift most days. As much as I'm embarrassed to admit it, I was an adrenaline junkie.

That's not to say I didn't appreciate flying for the pure sake of flying. I most certainly did. With the exception of training flights and check rides, neither of which I particularly enjoyed, I was always happy to strap myself into the cockpit. Of course, no pilot enjoys check rides, but there *are* those who claim to enjoy training flights. At the risk of sounding cavalier, I wasn't one of them. There was an earlier period in my flying career when I enjoyed training, but I was well past that stage by the time I joined STAR Flight. By then, I viewed training flights as a necessary, but extremely tedious, part of the job—something that should be kept to a minimum.

Unfortunately, the people for whom I worked subscribed to the theory that says if one training evolution is good, then fifteen of them must be great—especially when it came to rescue training. This seems perfectly logical at first glance, and if we had been playing golf for a living, I would have been on board with it. It's just that, ruling out lightning strikes or errant shots from other golfers, the likelihood of someone losing his life while grooving a swing on the practice tee is extremely remote.

The same can't be said of helicopter crews who continually rehearse hoist rescues and other high-risk evolutions. When you're conducting rescue operations, real or simulated, you're constantly inside that *dead man's curve* I keep talking about. It's not as if we were practicing with a net below us. The problem with practicing too many rescues is that it only takes one engine failure to kill you just as dead as if you were actually in the process of rescuing someone.

Don't get me wrong. I'm not saying that if your job is inherently dangerous, you shouldn't train for it. You still have to train, even if the training itself happens to be risky in nature. Nevertheless, it stands to reason that there's an inverse relationship between the risk factor associated with an activity and how often you should attempt to polish your skills by repeating it. As the task becomes more dangerous, the number of iterations required to reach the point of diminishing returns decreases accordingly. Once he has acquired the necessary skills, how many times does a rational person keep practicing a maneuver that is likely to kill him the very first time he screws it up?

Much to my disappointment, I was never able to sell the front-office folks on the merits of my argument. Thus, training (or *overtraining*, to be more precise) was perpetually one of those issues on which my superiors and I disagreed. And in management's defense, their rationale was not totally flawed. They believed they could improve the survival odds of their flight crews by more frequently exposing them to the lower relative danger of a controlled environment. That sounds reasonable, doesn't it? It's not as if they were trying to punish us by making us train. In their minds, they were simply looking out for our welfare.

Still, there's a fine line out there *somewhere* that defines that aforementioned point of diminishing returns. It would probably require someone with a degree in statistics to figure out exactly where that line exists. I can't even be sure who was right—or, more to the point, who (if anybody) was wrong—but my bosses and I definitely disagreed on the line's whereabouts.

At any rate, training flights and check rides aside, I still had one of the best jobs on the planet. And one of the principal ingredients that made being a STAR Flight pilot "one of the best jobs on the planet" was the compelling nature of the missions with which we were tasked. This is what I mean when I say it's impossible, in my mind anyway, to separate the *act* of flying a public-safety helicopter from the *reason* for flying it in the first place.

An Interested Observer

One of the things that made flying for STAR Flight so rewarding was the fact that when the pager went off, we had a clearly defined mission. Usually the task involved flying from the hospital, snatching up someone who was sick or injured, then flying them back to the hospital of origin as rapidly as possible. Occasionally, we would fly the patient, or patients, to a different hospital, but you get the general idea. Most of our time was devoted to treating patients and transporting them to a medical facility, just as the crew of a conventional ground ambulance would do. It's just that we got them there quicker and provided a higher level of care.

My job on these flights was simple. I transported the medical experts to the scene, occasionally fetched equipment out of the back of the helicopter, and then transported the experts back from the scene while they practiced emergency medicine in the back of my helicopter. I often told people that I was nothing more than an overpaid ambulance driver on EMS flights. It was the paramedics and nurses with whom I worked who were the key players. They were the ones who made the difference. Once we were on scene, I was pretty much relegated to the role of interested observer. It was in this role of "interested observer" that I witnessed a great deal of tragedy, and learning to cope with it was sometimes a challenging process.

During my first few years with STAR Flight, I tended to dwell on "bad" calls more than I should have. I would take a personal interest in the patients my crews and I had transported, often following up on them in the hospital. On the surface, this probably seems like a natural thing to do. But over time, it began to wear me down, just as it wears on the people who are actually providing care to the patients. In a way, I think it may even be worse when you're there as just a spectator.

It was precisely because I was *not* tasked with providing patient care that I was sometimes cognizant of the less-obvious details during a call, details that would understandably go unnoticed by someone who is actually working feverishly to save a patient's life. Unfortunately, especially as far as I was concerned, they were the kinds of details that made some calls difficult to forget.

"Sometimes, the scars are on the inside."

"What's the worst call you've ever run?"

When I was still flying, I got this question a lot. Usually, I would figure out a way to avoid giving an answer by changing the subject. Occasionally, I would simply pretend that I hadn't heard the question. Now that I'm retired, I think I can go ahead and answer it.

Of course, the predictable choice would be a call in which mayhem and carnage was the overriding theme. I certainly would have a number of those from which to pick. I still remember the day Dave Williams, the paramedic who flew with me on the night of my first call as a STAR Flight pilot, walked up behind me as we were preparing to leave the scene of a train-versus-pedestrian accident.

"Can you give me a hand here?" I heard him ask.

When I turned around, I was shocked to see Dave standing there awkwardly, and somewhat apologetically, with the vital signs monitor under one arm—and a severed leg under the other. After hesitating, just for a moment, I grabbed the monitor and loaded it into the aircraft. Then I took a disposable blanket from one of the cabin-mounted storage bins and handed it to Dave so he could wrap the leg before loading it into the back of our helicopter alongside the patient to whom it had formerly been attached.

There are a number of nightmarish calls that have left permanent abrasions deep behind my outer defenses, images that I wish I could purge from my memory. Obviously, the multiple-fatality car wreck I ran with Dave and Ron Startzel on the night of my first shift at STAR Flight is one of those. There are others as well.

The first time I transported a burn patient to BAMC (Brooke Army Medical Center), in San Antonio, I dutifully followed my crew inside the hospital. After we were in the elevator, headed up to the burn center, I wondered if I should have waited outside. Once we were wheeling our patient into the ward filled with all the other patients, some of whom had suffered horribly excruciating, disfiguring burns, I *knew* I should have waited outside. I made close to a hundred more flights to BAMC after that day—I never again set foot inside the burn center on any of them. I simply didn't possess the emotional fortitude to watch people agonize that way. It made me wonder—in awe, actually—how the people who cared for those patients had the psychological stamina to do it day after day.

After my experience at BAMC, I decided that burns had to be the worst form of trauma a person can suffer. Not that amputations weren't devastating injuries as well. Just like those burn patients I saw, they're not easy to erase from my consciousness—even though I witnessed some of them twenty years ago. I could fill an entire chapter recounting all the amputations I saw, but trust me, you wouldn't want to read it. There were also several decapitations I'd like to forget. I can think of one in particular—the result of a shotgun blast to the face—that has consistently haunted me in a way that I'm simply unable, or unwilling, to describe.

To be honest, regular exposure to that kind of trauma ultimately took a serious toll on my ability to relate to my "non-STAR Flight" friends and family. At the risk of sounding calloused, you almost have to file those kinds of grisly scenes away in an abstract part of your memory, or you'll succumb to the anxiety they produce. Even today, I choose not to recount those types of calls in graphic detail; but when I was still learning how to deal with the stress, I often came home and told my wife, Nancy, about the abysmal things I'd seen at work, thinking it would be therapeutic to talk about it—it wasn't. It certainly didn't do anything to improve Nancy's mental state, and since it really didn't help *me* to feel any better, I eventually quit communicating altogether, thereby allowing my inner demons to chip away unabated at my peace of mind.

As the years have passed, it's becoming a little easier to discuss some of the things I saw, and maybe writing about it here has finally helped me to exorcise some of those demons. Nancy has been encouraging me to write this book for a long time, and I even sat down and tried to do it about ten years ago, but the story I needed to tell was still a work-in-progress. Besides, I just didn't feel like reliving some of this stuff until recently. Now that I'm no longer accumulating baggage, it makes it easier to think about unloading some of it.

If this part of the book seems disjointed or out of place, I suppose it's because I'm writing it more for *my own* emotional well-being than for the benefit of the reader. In fact, I'm guessing some of you would prefer that I had left this discussion out entirely. If that's the case, I beg your indulgence. Perhaps I just decided to throw this chapter in here so I could get some things off my chest—which brings me back to answering the question at hand.

ANATOMY OF A STAR FLIGHT PILOT

The Worst Call I Ever Ran

Unfortunately, even though it happened over two decades ago, I still vividly recall most, if not all, of the details—those *peripheral* details I mentioned earlier. They're the kind of inconspicuous aggregates you tend to overlook unless you're standing around with nothing else to do while everyone around you is busy working. I hadn't been at STAR Flight for all that long, so I was still in the process of learning how to deal with the human-tragedy aspect of the job.

We flew to one of the surrounding counties directly east of Austin that night. When we landed behind a well-kept little ranch house, several emergency vehicles were already parked on the other side of the home, in the front yard. It was one of those picture-perfect nights—when the weather is balmy, and it feels good just to be outside. I remember this because after we had shut down, I decided to stay by the helicopter instead of going inside with the medical crew.

Like I said, I hadn't been at STAR Flight very long, and at that early stage in my career, I would normally have followed my crew into the house; but the weather that night was too perfect to waste, so I found a comfortable spot in the grass and just sat, soaking it in. We'd been dispatched to the house because someone inside had dialed 911 and informed the call-taker he was feeling chest pain. After we'd been there for about forty-five minutes, I remember thinking to myself, *This seems to be taking longer than usual.* Soon after that, I decided to go inside and see why they weren't bringing the patient out of the house.

As I entered the kitchen through the back door, the first person I encountered turned out to be the patient's wife, who looked to be in her late seventies or early eighties. I asked her if she could tell me where to find my medical crew, but she just gave me a blank stare. Then I heard voices coming from the other end of the house, so I headed off to see for myself what was taking so long.

I left the kitchen, then walked across a large living room and down a hallway. As I reached a bedroom door at the end of the long hallway, I found them administering CPR to a man who looked to be around the same age as the woman whom I'd met in the kitchen. They had moved him out of his bed and onto the floor because, in order for the chest compressions to be effective, the patient needed to be lying on a hard surface.

My medic was sweating profusely as he pumped the man's chest, and my nurse was on the phone with the ER doctor at the hospital. There were also a couple of ground medics there. One was busy preparing medications, and the other was confirming the dosages and taking notes for the official paperwork that would need to be filled out following the inevitable "pronouncement" from the ER doctor. I'd seen this drill enough to know they would continue performing CPR until there was no longer any hope, but based on the man's age, I also knew the ultimate outcome was fairly predictable.

Meanwhile, the woman who had met me at the door was all alone, so I decided to go back down the hall and check on her.

"Ma'am, are you okay?" I asked her.

Again, just as she'd done when I first entered the house, she looked right through me and didn't say a word. She kept wandering aimlessly, back and forth across the kitchen, and wouldn't even acknowledge my presence.

At first, I thought she was in a state of panic, but I finally realized she wasn't panicked at all. Instead, I'd been trying to converse with someone who was afflicted with Alzheimer's disease (or some other form of dementia), and I just hadn't been attentive enough to recognize it. Not only was she not panicked over her husband's situation—she was completely unaware of it.

I finally decided I was probably making her nervous, so I decided to leave her and walked back into the living room. As I looked around—because I had nothing else to do—I noticed some family photographs on the mantel above the fireplace.

There was one, in particular, that caught my attention, and I moved in for a closer look. Inside the aging wooden frame, I saw a strapping young first lieutenant in a World War II, Army Air Corps dress uniform. He was wearing a Tenth Air Force patch on his left shoulder—and a striking young blonde on his right shoulder. In the frame right next to that one, there was another photo. That same gorgeous blonde was locked arm in arm with a now-somewhat-older captain as they ducked beneath a shower of rice on their way out of a wedding chapel. The several rows of ribbons the young pilot was wearing beneath the wings on his uniform weren't in the first photo, but in this one, they were almost as conspicuous as the huge smile he was wearing on his face.

It was easy to deduce that the first picture had been taken before he was sent overseas, and the second was taken after his return. There must have been thousands of photos like these, all across America, taken during and after the Second World War— each of them telling a similar, yet uniquely personal, story.

On the opposite side of the room from the fireplace, there stood a bookcase. It was filled with more photos, and there were also several portraits, including one I knew I'd seen before. It was a reproduction of a very famous half-finished portrait—a portrait of President Franklin Delano Roosevelt. The portrait was unfinished because FDR, who was on a retreat to his summer White House in Warm Springs, Georgia, died before the artist could finish painting it. The remaining pictures were obviously family photographs, and from the dates on the photos and the ages of the people in them, I guessed that I was looking at three generations of people who could trace their lineage back to those faded black-and-white images on the mantel.

There were also several framed citations in the bookcase, which accounted for some of the ribbons the young captain was sporting in his wedding photo. I knew I was prying, of course, but I couldn't help myself. I had always been a big history buff, and as I've already mentioned—I had nothing else to do.

I started reading the citations and couldn't stop until I'd read every last word. They were quite compelling. Having flown C-47s with the Tenth Air Force, the young pilot had earned two Distinguished Flying Crosses and a Purple Heart while serving in the China-Burma-India Theater.

"Flying the hump" is what they called it. It was a massive airlift that lasted from 1942 through the end of the war, in 1945. It involved hundreds of planes, mostly C-47s, flying supplies from India to the Allied forces fighting the Japanese in China. These combat-resupply missions over the Himalayan Mountains were grueling and extremely hazardous. More than six hundred planes were lost, and more than a thousand crew members died in the effort.

Suddenly, I was overwhelmed with emotion. The man dying in the house that night—the man dying in the bedroom just down the hall from where I was standing—was an American hero. Because they were busy doing their jobs, however, not one of the people who were working in vain to *prolong* his life had any clue *about* his life—they had no idea who he was. What was truly tragic, though, was that the beautiful young bride in the wedding photo had no idea either. She and her husband were the quintessential couple from America's greatest generation; but at that particular instant, because I had no assigned task, I seemed to be the only person who realized that a national treasure was slipping away in the next room.

I can't begin to explain why that moment is so prominently magnified in my memory. I guess this just happened to be one of those situations to which I alluded earlier. It was one of those times when watching from the sidelines is actually harder than being on the field. It was hard for me to look at those photographs of a young, energetic man and his beautiful young bride—and accept that they were the same two people in the house that night. I suppose the fact that the dying man had been a military aviator, and a decorated one at that, made it even harder for me.

Still, having witnessed so many grisly scenes during my years as a STAR Flight pilot, it's simply not rational to think this could have been the one that affected me the most. So, why is it so difficult for me to talk about it? It's been more than two decades now, and I've never been able to articulate this story to another human being without losing my composure in the process. I've struggled with this so much over the years that I've simply quit telling the story because I'm not comfortable letting people see this chink in my emotional armor.

I'm a veteran, a hot-shot naval aviator. Crying is not supposed to be part of my genetic makeup. None of those gruesome calls that actually involved dismemberment and mutilation affected me this way. Those scenes may not be pleasant to remember, but I certainly don't break down in tears when I talk about them. How could this possibly have been the *worst* call I ever ran? Surely there's some rational explanation for it.

This is the paragraph where my journalism professors taught me that I'm supposed to cap off this chapter with a poignant answer to my own question. The only problem with that is, I *really don't know* why this call was worse than all the others.

It just was. It always has been. . . . It still is.

7

Flying without Fear
(The Early Years)

One of my all-time favorite movies is *Flight of the Phoenix*, the story of a rag-tag group of crash survivors who manage to rescue themselves from the Arabian Desert using their own ingenuity. Salvaging parts from the twin-engine plane they were traveling in at the time of the crash, they use them to fabricate a crude single-engine airplane and fly it to a remote oil field located miles from the crash site.

There is an early scene, prior to the crash, when Captain Frank Towns (Jimmy Stewart) is reminiscing with his navigator, Lou Moran (Richard Attenborough), about what it was like to be a pilot when he was young. As they sit in the cockpit of their Fairchild C-82, Frank Towns, with his hands resting casually on top of the yoke, turns to his navigator and delivers the following speech over the steady drone of two radial engines in the background:

> I don't know, Lou, . . . I suppose pilots are just as good now as they ever were, but they sure don't live the way we did.
>
> (dramatic pause)
>
> I can tell you that there were times when you took real pride in just *getting* there!
>
> (shaking his head wistfully as he continues)
>
> Flyin' used to be *fun*, Lou. It really did. . . . It was fun.

When I hired on at STAR Flight in 1992, the program was still in its infancy. We *did* take pride in "just getting there." And after we got there, we took pride in just getting back as well. Flying really was fun. Don't get me wrong. When I finally retired in 2012, I still enjoyed shaking the sticks, and flying was as rewarding as it had always been. It just wasn't as much fun. Thanks to things like GPS, night-vision goggles and state-of-the-art autopilots, the job of a pilot was definitely less demanding and arguably less risky—just not as much fun.

"Must we fly in this cracker box?"

The undersized Bell 206 I flew during my pre-employment check ride with Spanky Handley in 1992 was meagerly appointed compared to most present-day helicopters. Not only was it not equipped with a modern autopilot, it also lacked a flight control stabilization system, a feature now found on even the most basic of helicopters. Heck—forget autopilots and control stabilization.

There wasn't even a trim system in the thing. Instead, it was equipped with a rudimentary control-friction setup that could only be adjusted by turning a wheel, which was inconveniently located on the cockpit floor. After landing the helicopter, the pilot had to lean forward, reach down between his feet, then clumsily crank the wheel all the way to the stop—otherwise, the cyclic would fall over when he released it. The onboard avionics package consisted of a single communications radio and an archaic direction-finding navigation receiver, which looked as if it might have crossed the Atlantic with Charles Lindbergh.

Still, the little helicopter had a certain charm about it—that is, as long as you didn't have to transport patients in it. Weighing in at less than 5,000 pounds, it was mostly used as a training aircraft. We only flew it in service when the considerably larger and better-equipped Bell 412 was down for maintenance, which to our great consternation, occurred far more often than we would have liked.

Although the 206 was a fun little machine for the pilots to fly, the medical crews hated it—and for good reason. Unlike the 412, it wasn't equipped with a swiveling stretcher-base for loading and unloading patients. Nor was there a sliding cargo door. The 206 was seriously mis-engineered with a clumsy, double-hinged, swinging door, which only the most disturbed Rube Goldberg enthusiast could have appreciated. After opening the primary door, a second set of hinges had to be released with a pair of latches at the top and bottom of the door frame. Then, both sections of the unwieldy apparatus had to be folded forward and anchored to the airframe with a nylon strap. The strap was there to prevent the door from becoming a guillotine on windy days.

Once the door was out of the way, there was still precious little space in which to load the patient. Using a technique that closely resembled a Keystone Kops swinging-ladder routine, the medic, along with the nurse and at least one other person (usually a conscript from one of the ground units on scene), had to shoehorn the stretcher through the narrow opening and slide it forward until the patient's feet were in the cockpit, right next to the pilot.

This cozy arrangement whereby the patient's feet were in close proximity to the pilot proved to be problematic for me and my crew one day when, suddenly and without warning, I fell victim to an inflight attack from an unruly patient. Luckily for me, I happened to be flying with an experienced, well-trained flight medic. Paul Kuper was not only an excellent paramedic—he turned out to be a formidable wrestler as well.

"Please stop kicking my pilot in the head!"

Dispatched to a one-car rollover, my crew and I, flying the aforementioned and much-maligned Bell 206, flew to a winding country road about twenty minutes north of Austin. We arrived on scene to find a single patient, who upon being ejected from the vehicle, had suffered what appeared to be a severe head injury.

Every now and then, as any EMS professional can tell you, patients (especially those with head injuries) become agitated (sometimes even combative) when they're strapped down to a hard wooden backboard. This particular patient, however, was extremely well behaved and totally compliant, even during the rough and uncomfortable loading procedure. Of course, it's likely he was totally compliant for the simple reason that he was *totally unconscious* at the time.

The key point here is that Paul, in an effort to expedite our departure, decided to forgo affixing restraints to the patient's lower extremities. After all, the man was down for the count. To be fair, we *were* in a bit of a hurry. Our patient was barely clinging to life—or so Paul thought anyway. And in Paul's defense, he did ask me if I was okay with the seating arrangements before we lifted off.

Unfortunately, about five minutes into the return flight, the patient's condition greatly improved—and as previously noted, sometimes people strongly object when forced to lie perfectly still with their spine pressed against a piece of hardened lumber. Such was the case with this fellow, and since he couldn't get his arms free to punch Paul, he did the next best thing and began kicking the crew member who was closest to his feet (that would be yours truly) in the face. Now, I'm not usually one to complain, but I found this situation to be totally unacceptable, seeing as how I was trying to fly a helicopter, and the repeated blows to my head were making it *hard to concentrate!*

Recognizing that I was in some distress, Paul unstrapped and did his best to climb on top of the patient so he could grab his feet. Sadly, though, there simply wasn't enough room for one more person in the cockpit-sized arena where I was being pummeled mercilessly in a one-sided aerial kickboxing match. Acutely aware of the challenge I would soon face (mainly, trying to maintain control of the aircraft while comatose), I decided to make an unscheduled landing in a nearby farmer's field.

A minute later, we were on the ground, surrounded by a very impressive crop of chest-high corn. With the aircraft still running ("turning and burning," in rotary-wing parlance), Paul quickly unstrapped and exited the aircraft. Hacking his way through the menacing corn stalks, he eventually gained access to the cockpit through the copilot's door, whereupon he smartly secured the flailing patient's legs and feet. In less than two minutes, we were airborne and once again headed for the trauma center at Brackenridge Hospital.

The crisis had passed. Not only that—our bosses were none the wiser. Thanks to Paul Kuper's courageous actions (not to mention my ability to take a punch), a disaster had been narrowly averted, and all had ended well. Harkening back to Jimmy Stewart's soliloquy in *Flight of the Phoenix*—that scene where Frank Towns was waxing nostalgic to Lou Moran about a by-gone era when flying was fun—surely, our little inflight altercation had been the sort of extraordinary adventure about which he was reminiscing.

Experiences like this one were just part of the learning process as I settled into my new career as a STAR Flight pilot. Those first three years flying the line were an exciting time in my life. Even though I had honed my flying skills in the Navy, it was only after I started flying for Travis County that I truly began to mature as a pilot.

For one thing, I had always flown with a copilot in the Navy. Now I was the only aviator in a crew of three. There was nobody there to read a checklist to me—nobody there to keep me honest and help me to make critical decisions. And the situations that demanded a critical decision were more numerous than they had been in the Navy. Most of them involved weather. Deciding whether or not to launch in marginal weather conditions can be extremely stressful when lives hang in the balance. A lot of air ambulance operators make it a policy not to inform their pilots about the nature of the call when dispatching flights. That kind of policy suggests that the pilot, without regard for the patient's condition, should make his "go, no-go" decision based solely on the weather.

I couldn't disagree more. Fortunately, and I'm happy to give my former bosses credit for it, this wasn't the way we operated at STAR Flight. As experienced pilots, our management expected us to make responsible, common-sense decisions based on all the information we had at our disposal. If I was launching to transfer a stable patient from one hospital to another, it stood to reason that my personal weather minimums would be higher than if I was responding to an accident involving a busload of hemophiliacs. If dispatchers withhold those kinds of details, they do a disservice to the pilot; and if the pilot is not adequately informed before he decides to cancel a flight in marginal weather, they do a grave disservice to a critically injured patient as well.

Deciding whether or not to launch should be left to the aircraft commander. He needs to weigh all the relevant factors, including the input he receives from his medical crew. Again, this is one thing they always got right at STAR Flight. Not once during my twenty years did my superiors ever question my decision to cancel a flight for weather.

There *were* those times, however, when they occasionally asked me to explain why I had chosen *not* to cancel a flight in less-than-favorable conditions.

The Little White Room

When a pilot is flying in weather that permits him to see the ground and use it as his reference to remain upright, he's said to be flying in VMC (visual meteorological conditions). Conversely, a pilot is flying in IMC (instrument meteorological conditions) when he's unable to see the ground, in which case he has to rely on the attitude indicator in his cockpit as his sole visual reference. "Inadvertent IMC" is a term used in aviation vernacular to describe a scenario in which a pilot unintentionally flies into an area of poor visibility, usually clouds or fog, and unexpectedly loses his visual reference to the ground. Although it's not common, when inadvertent IMC does occur, it accounts for a large percentage of the deadly crashes that plague the air ambulance community.

There are at least a couple of reasons why this is true. Unfortunately, pilots who are not on instrument flight plans are sometimes reluctant to accept the reality that they have accidentally flown into IMC. Instead of immediately climbing to a safe altitude, which is what they should do, they think that if they can descend just a little, they might be able to regain visual contact with the ground. The problem here is that pilots who are operating in marginal weather ("scud running" we used to call it) are usually flying at lower-than-normal altitudes to begin with. As a consequence, *actual* contact with the ground too often occurs while trying to regain *visual* contact with the ground.

Another explanation for the crashes that occur following inadvertent IMC lies in the actual term itself. It's an *inadvertent* phenomenon, which means it's usually totally unexpected. This is especially true at night. If the pilot could actually see the cloud in front of him, it stands to reason that he wouldn't intentionally fly into it unless he was prepared to transition to instrument flight. When a pilot is flying in instrument meteorological conditions, he is literally sitting in a little white room. His visual reference to the rest of the world ends at that plexiglass canopy—the one located just a few feet in front of him.

This requires a totally different mindset. With no other means to remain upright and navigate, a pilot must fully commit to his instrument scan and totally ignore his external senses. There is no such thing as seat-of-the-pants instrument flying. It's an absolutely analytical undertaking.

This is normally not a problem for a well-trained, instrument-rated pilot who has filed an instrument flight plan and is mentally prepared to fly in instrument conditions. For whatever reason, however, that same well-trained, instrument-rated pilot, when he happens to be flying visually at 200 feet above the ground, will panic when he suddenly and unexpectedly finds himself surrounded by clouds. He doesn't panic because he's an incompetent pilot. He panics because he's human—a human who's in an extremely dangerous situation. It's what he does next that defines whether or not he's a competent pilot.

"If a picture is worth a thousand words, then the total lack of a picture is surely worth a few 'choice' words."

"Dammit! Dammit! Double dammit!"

That was the fancy pilot jargon I used to alert my crew that we had accidentally entered instrument flight conditions. We had been circling a residence for several minutes that night, waiting for the first responders to set up an LZ for us. I don't even remember why we'd been dispatched, but apparently I had decided that the nature of the call was serious enough to warrant launching in some pretty lousy weather. It was one of those mid-winter, Central Texas nights when the air is totally saturated, and the only thing that inhibits the formation of fog below the low-hanging clouds is the presence of a slight breeze.

Sure enough, while we were circling above the scene, I began to see the tell-tale reflection of our anti-collision lights flashing in the fog that was rapidly forming all around us. Suddenly, there I was—sitting in the little white room, along with Tom Bryan (my flight medic), who was riding up front in the cockpit with me. We were barely 200 feet above the ground at the time, and to make matters worse, I had been in the process of making a fairly steep turn when the ground disappeared below us.

Mad at myself for falling into the trap, I quickly began following the applicable emergency procedures for inadvertent IMC at low altitude:

1. Curse out loud and admit you screwed up.
2. Level the wings.
3. Increase power and establish a positive rate of climb.
4. Upon reaching a safe altitude, contact ATC and request an emergency IFR clearance to the nearest airport with a precision instrument approach.
5. Upon landing at said airport, call your boss (be prepared to eat some humble pie).
6. Thank your lucky stars you were in the Bell 412, which is fully rated for IFR flight. If you had been in the Bell 206, you probably wouldn't have made it past *Step 1* without augering in.

Before I could ask him to do it, Tom Bryan had already pulled out the approach plates, a book containing detailed diagrams of all the instrument approaches to nearby airports.

In preparation for just such an emergency, our medical crew members had been trained to read and interpret the approach procedures depicted in the book. They were also taught, just as a capable FAA-rated copilot would be expected to do, how and when to pass along the critical details of the procedure to the pilot during the approach. Because of the esoteric nature of the data and the complex manner in which it was presented on the approach plate, this wasn't exactly an easy skill to acquire. It was a responsibility normally reserved for trained, seasoned aviators.

Tom handled the task as if he'd done it a hundred times, talking me through the ILS (Instrument Landing System) approach to Robert Mueller Airport back in Austin. After breaking out of the clouds directly over the flashing approach lights at the end of the runway, we hover-taxied to the STAR Flight maintenance hangar.

Because the visibility was too poor to fly from the airport back to Brackenridge Hospital, that's where my fellow crew members spent the rest of the night sleeping on couches in the maintenance director's office. As usual, even though the weather wound up being unflyable for the remainder of the night, I stayed awake, just in case.

Before settling into the lounger in front of the television, I called Spanky Handley to let him know what had happened. By this time, Pat Leone had moved on, and Spanky had taken over as the director of operations. He was not at all happy and told me to meet him in his office the following morning. I figured he was mostly upset that I had interrupted his sleep, but this turned out not to be the case.

When I got off duty the next morning, I drove downtown and reported to the Travis County office building, as ordered. Spanky was waiting for me, his feet propped on his desk and his arms folded indignantly across his chest. Mike Phillips, who was still chief pilot, was waiting there as well.

When Spanky invited me to take a seat, I thought I was about to be lectured for making a poor decision to fly in marginal weather. Expecting to be admonished for having accepted a flight I should have turned down, I was surprised when Spanky threw me a curve ball. Instead of beating me up for launching in the first place, he began chewing me out for coming back IFR. He told me I should have descended until I could see the ground again, and then I should have landed and waited for the weather to improve. He went on to say that Travis County would probably be facing a fine from the FAA because I had returned on the ILS approach and STAR Flight was not certified for IFR operations.

Now, I had been fully prepared to take my medicine for launching when I probably shouldn't have, but this was something totally different. I admitted to Spanky that I had made a mistake by launching in the first place, but I adamantly defended the steps I'd taken once we were in the fog. I told him that I seriously doubted the FAA was going to fine us for following the emergency procedures they themselves had endorsed in our operations manual. I also told him it would have been reckless to descend after I was IMC.

Vehemently rejecting my argument, Spanky continued to berate me for making the decision to climb instead of descend. Then he finished by warning me to never do it again.

"Next time, just land!" he scolded me.

As I sat there in disbelief at what I was hearing, I turned to Mike Phillips. I could tell he was almost as incredulous as I was, but it wasn't until he began to speak that I was sure his incredulity was directed at Spanky instead of me.

"Spanky, he did the only thing he could do," Mike said in my defense.

Apparently, Mike had thought Spanky was going to deliver the lecture I was *expecting* instead of the one I got. To my relief and Mike's credit, he took my side in the argument. He told Spanky that, once I had flown into the fog, I had *absolutely* done the right thing in climbing instead of descending, especially since I was only a few hundred feet off the deck at the time.

Still, Spanky kept insisting emphatically that "we are *not* an IFR operation" and repeatedly told me to "just land" next time. Even though I was still the newest and youngest STAR Flight pilot, I respectfully told Spanky that, if I ever found myself in that situation again, I was going to do exactly the same thing I had done this time.

As it turned out, by the time I *did* find myself in that situation again, Spanky was no longer the director of operations. Fortunately for me, his successor was much happier with my decision to come back IFR than he would have been if I'd followed Spanky's advice instead.

The "just land" dressing-down from Spanky was only the first of many lectures I received during my career at STAR Flight. For some reason, maybe it was my propensity to say whatever I happened to be thinking without regard for the consequences, I had a hard time staying out of trouble. Sometimes, as was the case following the inadvertent IMC incident, the reprimand was arguably unjustified (actually, that one was *completely* unjustified). Then there were other times, especially early on, when I positively had it coming to me—like the day I decided to show just how good I was at squeezing a large-sized helicopter into a small-sized landing zone.

If the LZ's not big enough when we get there, it will be by the time we leave.

—Yours Truly

South Austin is full of quaint residential neighborhoods, the kind where June Cleaver (if you can imagine her living there instead of on the set at Universal Studios) could feel safe sending Wally and the Beaver out the door to school every morning. On this particular day, however, Wally and the Beaver would be in for some unexpected excitement in their quiet little neighborhood. You see, *Ward* Cleaver had called 911, complaining of chest pain, and the STAR Flight helicopter that was coming to their father's aid was about to put on quite a show, transforming their tranquil neighborhood into a scene of unexpected chaos.

The street on which the Cleavers lived was heavily wooded, and the firefighters who had preceded us to their residence radioed that the closest available landing spot was in the parking lot at Beaver's elementary school, about a half mile away. The on-scene commander said he would send two of his men to set up an LZ at the school, and from there, they would use their truck to transport my medical crew and their equipment back to the Cleaver residence. This seemed a little inconvenient to me, and as we arrived, I spotted what I thought would make a nice LZ through a small opening in the trees on either side of the street in front of the house.

"Tell them I can get it in there," I said to my medic, who was handling the radio communications with the firefighters.

This was going to be a superior display of airmanship, the likes of which these young firefighters and the gathering crowd of onlookers had never before seen.

As soon as they had secured the street, I began my approach. It was a thing of beauty as I expertly hovered the Bell 412 a few feet above the tops of the majestic pecan trees, their supple branches spreading apart in my rotor wash as if they were yielding to my nobility and beckoning me into their domain. Then, as the spectators watched in amazement, I artfully slipped my powerful, heavier-than-air flying machine through the narrow gap in the canopy with mere inches to spare. Once the rotor was beneath the branches that had been guarding the entrance to my LZ, I expertly guided my royal ship forward to avoid a parked car and gracefully set it down on the street below.

I was quite certain that, had he witnessed it, Igor Sikorsky himself would have been impressed at the skill with which I had just landed my 12,000-pound helicopter in this congested, densely wooded South Austin neighborhood. There were no fewer than three mailboxes and one street sign neatly tucked beneath the Bell 412's forty-six-foot main rotor.

The medical crew was already on their way to check on Ward Cleaver when I shut the engines down and applied the rotor brake. As I opened the pilot-side door, the on-scene commander hopped up on the skid-mounted step next to the cockpit and shook my hand.

"That was the damnedest thing I've ever seen," he said, smiling. "I would never have guessed you could get this thing in here."

I thanked my admirer for his kind words, and gracefully descended from my cockpit to survey the scene of my great and awe-inspiring accomplishment. I have to say, even as I look back on it now, it was some very impressive stick work.

Regrettably, however, getting out would prove to be much harder than getting in. Even as I was waiting for my crew to bring the patient out of the house, I started to become uneasy about the situation in which I now found myself—and it didn't take a PhD in Newtonian physics to understand why. As anyone with even a basic understanding of helicopter aerodynamics is aware, the vortex created at the tip of a turning rotor blade flows down and out, then up and in. The same rotor wash that was blowing the tree branches away from my helicopter during our descent through the canopy was going to be sucking them toward it on the trip back up. I had a really bad feeling that those "mere inches" I had to spare coming into this tiny LZ were going to vanish on my way out.

The good news was, Ward Cleaver's chest pain turned out to be nothing more serious than indigestion. Now, the only suspense left on this call was waiting to find out if my exit from this majestic outdoor stage would be as graceful as my entrance.

I greatly feared it would not be.

Unfortunately for me and my throng of adoring fans, which had grown even larger since our spectacular entrance, my instincts were right-on. Our departure from the previously well-manicured street was not nearly the crowd-pleaser that our arrival had been.

Although we didn't sustain any damage to our helicopter, the same could not be said for the Cleavers' brand new Chevy Suburban parked below us. As luck would have it, one of the more sizeable limbs we severed during our exodus went crashing down on top of it, cracking the windshield and producing a good-sized dent in the hood. Of course, I couldn't see that from my vantage point in the pilot's seat. I did, however, see several spectators ducking for cover as they dodged shrapnel from the less-substantial lumber we were chopping on our ascent through the trees—not to mention the barrage of almost-ripe pecans, which we were turning into high-speed projectiles with the near-supersonic tips of our main rotor blades.

Although there were no reported injuries, my poor decision that day eventually ended up costing the Travis County taxpayers a little over two-thousand dollars in property damage. Oddly enough, there are some aviators who might say this was a small price to pay for a supremely exciting demonstration of the helicopter's unique and amazing capabilities. Sadly, though, my boss was not one of them, and much to my chagrin, those were not the words he chose to describe the incident.

Today's STAR Flight pilot would likely be suspended or terminated for doing what I had done. Back then, it was barely noteworthy. Of the five who were already around when I was hired by Travis County, there wasn't a single pilot who didn't use his helicopter as a pruning saw at some point during his tenure there. It was nearly a dozen years and several scrapped rotor blades later before anyone actually lost his job over it.

Fortunately for me, this particular lapse in judgment had occurred back during the early years—when flying was still fun.

Our Bell 206-L with the "Rube Goldberg" door prominently displayed.

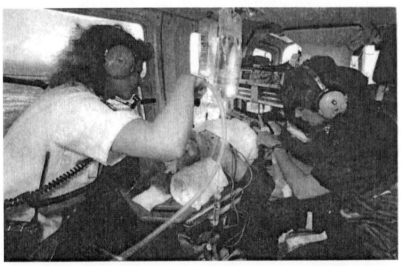

The tight confines of the 206 interior made the diminutive helicopter extremely unpopular with the STAR Flight medical crews.

Our Bell 412, weighing in at 11,900 pounds (max gross weight), was much more capable and nearly three times heavier than its little brother.

The Bell 412 on short final to the Brackenridge Hospital helipad with the Texas Capitol dome in the background.

A family photo, taken in front of the Bell 412 in 1993—My son, James, was not quite as camera shy by this time, but neither he nor his older sister, Lee Ann, look all that excited to be posing in front of Dad's helicopter.

STAR Flight circa 1996—Spanky Handley is sitting atop the helicopter (left), directly above yours truly.

8

Flying in Exile
(The Admin Years)

During my first three years at STAR Flight, 1992 to 1995, we had to complete a manually calculated weight and balance form for every flight on which we transported a patient. The FAA required the pilot to calculate the aircraft's takeoff weight and center of gravity each time we launched with passengers aboard—and according to the FAA, unconscious patients were no different from people who buy tickets to board airliners. The net result was that I often found myself turning and burning in the middle of a highway, crunching numbers with my head down, while everyone else was running around with their hair on fire.

It doesn't take much imagination to see how this could be less than safe. Common helicopter protocol dictates that nobody comes or goes under spinning rotor blades without an acknowledgment (thumbs-up) from the pilot at the controls. Well, if the pilot has his head buried in the cockpit instead of looking outside, this can be problematic.

After three years of this nonsense, I thought to myself, *There has to be a better way*, so I set about trying to come up with a system that would satisfy the FAA and still allow us, as STAR Flight pilots, to maintain our situational awareness when we were hot-loading patients on a scene. It was called a "load schedule," a series of handy charts the pilot could reference to complete all of the FAA-required calculations without the aid of a slide rule (we actually had a calculator—I just threw the "slide rule" in there to make it sound better).

Pat Leone, to his credit, had come up with an elementary load schedule back when he was the director of operations, but it was only a first step toward solving the problem. It helped in calculating the aircraft's center of gravity, but it still required the pilot to determine his takeoff weight manually. These manual calculations had to be accomplished using the current fuel load and the actual weights of each occupant aboard the aircraft—even the patients. The pilot was expected to ascertain the patients' weights by whatever means he could.

Of course, if the patients were able to communicate, you could always ask them how much they weighed, but if they were unconscious (or even in severe distress), you can see how this might be awkward. In order for a load schedule to be truly beneficial, the pilot not only needed to be relieved of the need to determine patient weights, he needed to be freed from the burden of using a calculator altogether.

The ultimate answer to the problem lay in convincing the FAA to let STAR Flight use a tool called "weight averaging" in our load schedule calculations. After submitting several weight-averaged load schedules to our FAA District Office, I was finally able to produce one that met with their approval. The end product comprised more than forty pages of spreadsheet calculations that read like sheer poetry to those of us who no longer had to worry about passing a trigonometry exam every time we flew. Now, instead of fumbling around with a calculator, pen, and scratch paper, the pilot could satisfy the FAA requirements simply by reading the pre-calculated numbers from a chart.

So, life at STAR Flight was good.

Like most people from my generation, I was still learning my way around a personal computer back in 1995. And like a lot of my peers, when I did spend time in front of my computer, I was mainly using it as a high-dollar word processor. However, during the process of developing our new and improved load schedule, I became fairly proficient with spreadsheets. Soon after, I began to play around with database software as well.

Not long after my bosses and I had convinced the FAA to accept our automated load schedule, I began working on a comprehensive STAR Flight database that consolidated all of our maintenance, operational, and medical data. It took me three months to build, but once the database was operational, we used it to produce everything from billing statements to FAA-mandated maintenance forms. This went a long way toward streamlining the administrative duties in both our front office and in our maintenance department. It also lessened the amount of paperwork the pilots and medical crews had to complete after each flight.

Now, life at STAR Flight was even better.

LIFE INSIDE THE DEAD MAN'S CURVE

No good deed goes unpunished.

—Oscar Wilde (author)

It wasn't as if I had set out to become STAR Flight's resident mainframe geek. Tinkering with the computer was mostly just a tool I had employed to stay awake when I was on night shifts, but it turned out I was reasonably adept at building spreadsheets and databases, and for whatever reason, I seemed to be the only person at STAR Flight who had any interest in pursuing those types of projects.

In selling our load schedule to the FAA and designing our fully automated database, I had raised my stock in the eyes of my bosses. So what did I get in return for all that extra effort? Spanky Handley offered me the chief pilot's job—and lest you think this was necessarily a good thing, the only reason the position was available was because Mike Phillips, the current occupant at the time, desperately wanted out. He and Spanky, who was still the director of operations, didn't always see eye to eye, and they had mutually decided that it might be a good idea if Mike and I traded jobs.

Although it wasn't nefarious at all, Mike also had an ulterior motive. Just like most of the pilots at STAR Flight, yours truly included, Mike Phillips enjoyed flying a helicopter a lot more than flying a desk, and he felt like he had already served his time in exile. Now he wanted to move back to the line, and after all Mike had done for me, I figured I at least owed him this favor in return. Besides, this was technically a promotion, and I suppose I should have been flattered that I was being offered the position over the more senior pilots. It's also true that the promotion to chief pilot came with a corresponding pay increase. It's just that I was perfectly happy with the way things were. I liked being a line pilot. I hadn't taken this job hoping to climb my way up the STAR Flight corporate ladder.

Still, Mike had campaigned for me during my hiring process, and I did owe him a favor, so I reluctantly agreed to take the job under the condition that I would be able to continue flying on a regular basis. So, at the ripe old age of thirty-nine, barely three years into my career at STAR Flight, I gave up my line-pilot position (a job for which I had worked so hard) to become Travis County's newest chief pilot.

This was a big adjustment for someone who had been living the dream up to this point. Those early years account for some of my favorite memories at STAR Flight. I'm guessing there aren't that many people in this world who look forward to going to work every day, but I actually did. Now that I held the title of chief pilot, things were going to be different. From now on, on most of the days I came to work, I was no longer going to be sitting in the cockpit of a helicopter. Instead, I was going to be sitting behind a desk, working inside an office—just like people who had real jobs.

Don't get me wrong. My job as chief pilot wasn't exactly a miserable existence. Spanky did make good on his promise to let me fly, though it wasn't as often as I would have liked. He and I were responsible for covering vacation shifts, and in an effort to keep me happy, he allowed me to take the lion's share of those shifts for myself. The thing is, I really didn't think this constituted much of a sacrifice on his part, seeing as how I wasn't sure Spanky enjoyed working regular shifts all that much anyway. Still, that didn't really matter. I was just glad to be flying when I could.

Even though I was thankful to be flying most of the vacation shifts, a conflict soon arose over the matter of who was going to train the line pilots. When it came to training flights, as well as post-maintenance check flights, we routinely disagreed over who should go flying and who should stay behind in the office.

When Spanky had replaced Pat Leone as the director of operations, he'd elected to retain his status as the program's instructor pilot. It was certainly within his purview to do so, but that's a job normally reserved for the chief pilot in most air ambulance programs. Of course, there was a reason Spanky had become STAR Flight's designated flight instructor back when he was only a line pilot. He was exceptionally good at training pilots. This, coupled with the fact that he enjoyed it and wanted to do it himself, left the two of us with a problem. Having spent three years as a Navy flight instructor, I was good at training pilots too—and I was the chief pilot, the guy who was actually *supposed* to be doing it.

The support for my argument, as far as I was concerned, was in the official job descriptions. Although the FAA mandated that STAR Flight's director of operations had to hold a commercial pilot's certificate, he wasn't required to stay current or, for that matter, even to be qualified in any of the aircraft that we were flying at the time. He wasn't even required to hold a current medical certificate. Conversely, the regulations specifically obligated the chief pilot to meet each and every one of these requirements. The chief pilot had to be able to fly. That's why he was called the chief *pilot*. Nowhere in the regulations was there a requirement for the director of operations to ever leave his desk.

I've always maintained that, under the FAA's regulations, the director of operations is the easiest guy to replace in a commercial aviation outfit. You can bring in his successor from outside your organization, and as far as the FAA is concerned, he can start acting in that role the same day you hire him (no training required). That's because you're not hiring him to be a pilot. It's supposed to be an administrative position.

I guess I didn't blame Spanky for wanting to have it both ways. If you could get away with it, why wouldn't you want to be the guy who gets to run the airline and the guy who actually gets to fly the planes too? At that time, the next person above Spanky in our chain of command was the Travis County EMS director. He worked in a downtown office, far removed from STAR Flight. He didn't know much about our day-to-day operations, and he knew even less about FAA regulations. He certainly wasn't going to issue an ultimatum to Spanky Handley, who had been there since the beginning of the program, to decide whether he wanted to be the director of operations or the chief pilot.

I quite often made that very argument to Spanky—but it didn't carry much weight coming from me.

A New Sheriff in Town

In late 1999, roughly four years after I had become chief pilot, Travis County got together with the City of Austin and decided to jointly hire someone to oversee both the aviation and medical sides of the operation. They appointed Casey Ping, who was the chief flight medic at that time, to be the first STAR Flight program manager. Casey's work ethic, along with his expertise in the area of rescue operations, quickly paid dividends after he was put in charge of the program. The "R" in STAR Flight stood for rescue, and Casey Ping, more than anyone else, was the reason we became experts at it.

More importantly, at least with regard to his meteoric rise within the organization, Casey was an expert politician. And like most politicians, in addition to his proponents, he had his share of detractors as well. Both camps were passionate when it came to their respective opinions about the man who was now running the show at STAR Flight.

And to which camp did yours truly belong? Well, naturally, I belonged to both. On any given day, my opinion of Casey Ping largely depended on how long it had been since he and I had last engaged in one of our many arguments.

We had some good ones, too—the kind that could have gotten both of us fired if they hadn't taken place behind closed doors. Occasionally, because of the volume levels involved, even the closed doors and insulated walls weren't sufficient to keep our heated conversations private. I guess you could say we were both passionate about our respective opinions. Through it all, I never questioned Casey's intentions. We both wanted to make STAR Flight a better program. It's just that we sometimes disagreed, to put it mildly, on how to go about doing that. Still, I had to respect his commitment to the cause, and I think he respected mine as well.

If I had to make a baseball analogy (and I don't—but I will), I would say Casey was the "Billy Martin" of bosses. When Billy Martin played for the Yankees, some of his teammates absolutely loved him, but because he was so passionate and intense, others didn't even want to share the same dugout with him. Later, when he became a Major League manager, it was the same story. Some of his players thought he was the greatest manager since Casey Stengel, while others said they couldn't play for the man and asked to be traded elsewhere. There was one thing about Billy Martin, though—everywhere he went, he made things happen.

That was Casey Ping. The energy he brought with him to STAR Flight's front office was unprecedented. He made his impact felt immediately, and the results were nothing short of remarkable. He took an already-good air ambulance outfit and, in the short span of just a few years, transformed it into the foremost public-safety helicopter program in the United States.

Not long after Casey had become STAR Flight's first program manager, Brackenridge Hospital withdrew from the STAR Flight management triad, and the nurses joined the medics as City of Austin employees. In the years that followed, the city would eventually cede total control of the program to the county, which meant that, instead of three different employers supplying crew members to staff the helicopters, everyone at STAR Flight worked for Travis County. This unification in our chain of command had been long overdue, but the process was not entirely without complications.

There were a few people within STAR Flight who struggled when it came to shifting their allegiance to a new administration. Moreover, a number of the medical crew members were long-time city employees, and they already had vested retirements with their old employer. This mainly affected the paramedics. The nurses had only been employed by the City of Austin for several years, and the majority of them, unlike most of the medics, didn't have a huge stake invested in their city retirements. In the end, most of the experienced crew members elected to remain with STAR Flight, which is a testament to the commitment they shared to their professions and to the program as a whole.

None of this had much effect on the pilots. Nor did it have much of an effect on the working relationship between Spanky and me. I was still plugging away on administrative tasks that I felt should have been Spanky's responsibility, and he was still taking many of the flight hours I thought should have been mine. Unable to resolve the issue on our own, Spanky and I went to Casey one day, looking for him to fix the problem for us. Instead, he told us we were both acting like little kids and unceremoniously threw us out of his office.

He was right, of course. Spanky and I *were* little kids. That wasn't necessarily a bad thing, though. One of the primary reasons I had become a pilot was so I wouldn't have to grow up and get a regular job like other adults. I don't think I'm going out on a limb when I say the same was probably true for Spanky Handley. We were very much alike when it came to our passion for flying, which was why we were butting heads in the first place.

Unable to gain any resolution with regard to our respective job descriptions, Spanky and I continued marching, and for the most part, we both did a fairly decent job working around our differences. This was arguably easier for Spanky than it was for me. After all, he was happy with the way things were. I was the one who was looking to alter the status quo.

Still, even though there were many days when I wished I had remained a line pilot, I really had it pretty good. Because my office was located at the hangar, I was able to spend a lot of time hanging out with the mechanics, who taught me a lot about the various systems on our helicopters. As a pilot, I found this to be useful knowledge because it helped me to better understand the principles behind many of the operating procedures contained in the flight manual, especially the emergency procedures.

Actually, referring to *aircraft maintenance technicians* as "mechanics" doesn't do them justice. Just like pilots, they have to be certified by the FAA, and this is done only after countless hours of specialized training and testing. Stan Wedell, who would later be promoted to director of maintenance, had been at STAR Flight almost as long as I had. Stan and I had established a good rapport even before I became chief pilot, and after I moved to the hangar, we became good friends. He shared an office with Mike Self, who came to STAR Flight in 1997 and was a major asset to the program from the time he arrived. Even though Mike had jumped through all the hoops to become an aircraft maintenance technician, he always introduced himself as "Mike the mechanic," a tongue-in-cheek reference to the rock group Mike *and* the Mechanics. Both Stan and Mike had learned their trade in the military. Stan had served in the Air Force, and Mike had cut his teeth maintaining AH-1 *Cobras* for the Army.

Those guys were definitely the unsung heroes at STAR Flight. As pilots, we did our best to break helicopters as fast as Stan and Mike could fix them, but we were never out of service for lack of an aircraft. Because STAR Flight was a twenty-four-hour, seven-day-a-week operation, at least one maintenance technician had to be on call around the clock. This meant that the standby technician, after he had already worked a full shift during the day, would often be called in to service an aircraft in the middle of the night. Then, even though he might only get a few hours of sleep, he was expected to show up bright and early the next morning to start it all again.

Through it all, I never heard either of those two guys complain about being overworked. Not only that, Mike was my "go-to" expert whenever I had questions during my do-it-yourself automotive projects. When we were together at the hangar, he would never let me work on the helicopters, but he was always happy to give me plenty of professional advice on how to turn wrenches on my own vehicles. There were other good maintenance technicians who came and went during my career at STAR Flight, but none served longer, and none were more respected than Stan Wedell and Mike Self.

When I wasn't hiding out with Stan and Mike, I would occasionally escape from my office by driving over to the Brackenridge crew quarters for unscheduled training flights with the line pilots. These weren't "hot seat" sessions by any stretch of the imagination. I think most of the crews actually enjoyed them, especially the pilots. This was probably because, after we had finished training, I would normally send the on-duty pilot home and work the remainder of his shift myself. The pilots were salaried employees, so they were basically being compensated with free vacation time in return for letting me subject them to my spontaneous teaching sessions.

Unfortunately, someone in the county's morale-suppression department eventually caught wind of the free vacation time, and I was forced to end the practice of relieving the on-duty pilot. This didn't preclude me from taking his place in the cockpit, however, so instead of sending the pilot home after we trained, I would simply leave him behind at the crew quarters while I flew all the calls that came in for the remainder of his shift. Instead of being compensated with free vacation time, the pilot was rewarded with uninterrupted nap time. Because he was technically still at work, we were able to circumvent the wet-blanket "no free vacation" edict without ruffling any bureaucratic feathers.

Sometimes I would take my good friend, Jim Allday, along with me on these impromptu training exercises. Jim, you'll recall, was the guy who had answered the door to the STAR Flight crew quarters years earlier, when I had first shown up there looking for a job. He was "Mr. STAR Flight." As the last remaining original crew member, Jim had been elevated to the position of clinical manager, and just as I now was, he had been incarcerated behind a desk at the hangar for a couple of years.

I think Jim was just as happy to get out of his office as I was to get out of mine. In fact, he and I flew together quite often during my tenure as chief pilot. Jim was an interesting personality, and for that reason alone, many of the calls he and I ran together proved to be interesting as well.

CSI Travis County, Texas

One of the great things about flying a helicopter, especially a public-safety helicopter, is the scenery. If you're at 30,000 feet and 500 knots, flying in a jet, everything below you looks pretty much the same. Not so if you're in a helicopter, flying at 200 feet and 140 knots. Not only do you get to enjoy a bird's-eye view of the terrain, you actually get to observe people going about their daily lives. If you're over the city of Austin, for example, you can even see which streets are flowing and which ones aren't. If there is a traffic jam, it's usually not hard to figure out why.

It's too bad people who stubbornly camp in the fast lane while driving a few miles per hour below the speed limit can't see what a mess they cause on crowded roadways. If I were a police officer, I would make it my pet project to ticket as many of these inconsiderate knuckleheads as I could.

As you move out of the city and into the suburbs, you can't help but notice how many people own swimming pools. By the way, if you're ever in a position where you're thinking about purchasing a pool, you should know that it's been my observation that people who own them don't actually swim in their pools. Occasionally, you might see someone lying next to one, but after looking down at thousands of swimming pools in my career, I don't think I ever truly saw anyone swimming in one. Now, don't get me wrong. It's certainly not my intent to dissuade people who think they need a swimming pool from buying one, but if you're only going to use it as an excuse to soak up some rays, I'd recommend just buying some pool furniture instead. It's far less expensive and much easier to maintain.

Finally, as you move out farther, past the suburbs and into the rural areas, you begin to appreciate just how expansive the planet actually is. Much of it is virtually uninhabited. So many places are inaccessible by motorized vehicles, that it's hard to believe you could ever find anyone who truly makes up his mind he doesn't want to be found. Even near population centers, there are always places off the beaten trail where people can go to disappear from the rest of society.

This is definitely true in Travis County. The rugged hill-country terrain west of town offers plenty of good hiding places, places where you would most likely be able to avoid detection from people intent on finding you. Still, just because it's *difficult* to find a needle in a haystack, it doesn't necessarily mean it's impossible. Sometimes you just stumble onto things you weren't even looking to find. This was certainly the case one day when Jim Allday and I were flying to a call on the western edge of the county.

Dispatched to a chest-pain call in a small community about thirty miles west of Austin, we were barely outside the city limits when I happened to look down through the trees and caught a glimpse of something that seemed very odd. Without saying anything to warn my crew, I immediately wrapped the aircraft into a ninety-degree angle-of-bank turn to go back and take another look at what I'd seen.

Jim wasn't pleased with the unannounced course reversal. Neither was Stephen "Stef" Maier, the other medical crew member flying with us that day. Cruising at 140 knots, this is a fairly violent maneuver, but I needed to make sure I didn't lose sight of the tiny clearing where I'd seen whatever it was that had gotten my attention. Don't forget, we were already responding to an emergency, so it wasn't as if I had a lot of time to waste searching for something when I wasn't even sure why I was trying to find it.

Stef was quick to remind me of this fact. By all rights, I should have kept flying straight toward our assigned destination. Neither Jim nor Stef were happy with my decision to turn around, but I convinced Jim to get on the radio and tell our dispatcher that we needed to delay long enough to figure this thing out. As he acknowledged our transmission, the skepticism in the dispatcher's voice was clearly palpable.

"This had better be good!" Jim exclaimed from the copilot seat.

At just that instant, I saw what it was that had caught my attention.

"Well, then you're not gonna like it," I said to Jim. "I'm pretty sure this is *not* gonna be good."

Then Jim saw it too.

It was a body—an adult male—lying prone on the white limestone rocks below. This was a typical scorching-hot, midsummer day in Central Texas, so the rocks on which the man was lying had to be hot enough to fry an egg. This was obviously not a hiker taking an afternoon nap.

As we circled around and got lower to take a closer look, we could see that the man's head was resting in a pool of blood, and there was a handgun on the ground near the body. We described what we were seeing to the dispatcher and asked him if we should continue to our original call or land and investigate the scene below us. After a short pause, we were taken off our original call and reassigned to the one upon which we had just stumbled.

This is when Jim Allday began his amazing metamorphosis into a professional crime-scene investigator. A huge television and movie enthusiast, Jim was a big fan of *CSI*, the popular television crime-drama that revolved around a team of forensic investigators in the Las Vegas Police Department. After declaring the area a crime scene while we were still circling 200 feet above it, Jim announced that we had to land far enough away to avoid disturbing any evidence with our rotor wash.

"Shouldn't we try to make sure the guy is actually dead before we start the murder investigation?" Stef asked sarcastically. "We might actually be able to keep him alive if he doesn't bleed out waiting for us to get to him."

As it turned out, Stef Maier's contention that we should land in close proximity to the body didn't much matter. The clearing was barely big enough to accommodate our helicopter—and the man was lying right in the middle of it. This would have required me to land directly on top of him and straddle his body with our skids; and aside from the fact that Jim would have been furious at me for disturbing his crime scene, this didn't strike me as a particularly good idea. The next closest clearing was more than a quarter mile away, so much to Jim's relief, we were able to land without compromising any of his crime-scene evidence.

In the meantime, our dispatcher had relayed our position to the Travis County Sheriff's Office, and a couple of deputies were already enroute. Because there were no nearby streets, however, it was going to take them a while to get there, which suited Jim just fine.

It took us a few minutes to hike to the scene, and just as Jim had suspected, the man had been dead for at least several hours. Stef radioed our dispatcher to let him know we had an obvious DOS from an apparent self-inflicted gunshot wound to the head. Jim was having none of that, however, as he immediately jumped on the radio to inform the dispatcher that we couldn't be sure the wound was self-inflicted and directed him to inform the sheriff's deputies that we were treating this as a potential crime scene.

Stef looked at me and rolled his eyes in frustration. He was understandably displeased that his partner had rebuked him over the radio for all the other EMS units to hear. Then, to make matters worse, Jim began lecturing Stef.

"How can you be so sure this is a suicide?" Jim asked.

"Oh, I don't know. . . . That revolver lying next to his hand might be a pretty good indicator," Stef retorted.

"The killer could have left it there."

"That's right, Jim. He left the murder weapon there so the cops would be sure to find it. It's probably just his way of taunting them, you think?"

The next twenty minutes were as entertaining as they possibly could have been under the somber circumstances. As Jim walked around and meticulously searched for clues, Stef continued to mock him unrelentingly. When Jim countered, it was like watching an old married couple bicker back and forth at one another. Finally, the two Travis County sheriff's deputies arrived, at which point Jim, having not yet concluded his own investigation, began supervising theirs as well.

After several minutes, one of the deputies politely informed Jim, "We can handle it from here."

Shortly thereafter, they invited us to leave, much to Stef's amusement.

For the rest of that day, Stef continued to give Jim a hard time over his amateur sleuthing. The funny thing was, it didn't seem to bother Jim in the slightest. It was classic Jim Allday. He knew how to roll with it as well as anybody I've ever known.

I learned this myself when I mercilessly teased him one morning after he had called in to a local radio talk show the previous evening. The host of the talk show was interviewing a Hollywood casting director, who was in Austin with the producers of *Friday Night Lights*, the movie that follows the Odessa Permian football team during their run to the Texas high school championship game in 1988. I happened to be listening on my drive home from work when the host took a caller whose voice I instantly recognized. It was Jim Allday.

The producers were looking for extras who had played football in high school, and Jim, who *had* played high school football and had even played *at* Permian (albeit a quarter century ago), began making his case for why he would be a perfect fit for their movie. I couldn't help but laugh as I listened to Jim make his shameless pitch to become a Hollywood star, only to be told he didn't exactly fit the demographic they were hoping to attract.

As soon as I got to work the next morning, I walked straight into his office and began giving him a ration of grief over his pathetic phone call to the radio station.

"What's wrong, Jim? The casting director wasn't interested in putting a forty-year-old fullback in his high school football movie?"

At first, Jim kept a straight face and pretended not to know what I was talking about, but then his theatrical look of bewilderment gave way to a sheepish grin as he confessed to making the call.

Jim's good-natured willingness to be the butt of the joke, especially if it made the other guy's day, was a rare quality among the type-A personalities at STAR Flight. Even though he was the senior-most crew member in the program, he was more than happy to mix it up with even the newest of rookies on equal terms. Unlike many of our fellow crew members, I tried not to take advantage of Jim's even-keeled personality unless the opportunity was absolutely teed up in front of me. Unfortunately for Jim, his passion for the silver screen presented me with just such an opportunity one night when we delivered a patient to Seton Hospital in downtown Austin.

Jim had just wheeled our patient into the emergency room, and I was securing the clamshell doors on the back of the helicopter. I noticed a man, not far from the helipad, smoking a cigarette and watching me with great interest. I immediately recognized the curious onlooker. It was Dennis Quaid. I had read Tom Wolfe's book about the early days of America's space program back when I was in college, and *The Right Stuff* was later released as a motion picture at about the same time I graduated from Navy flight school. Quaid's portrayal of Gordon Cooper, one of the original Mercury astronauts, was one of the things I enjoyed most about one of the greatest aviation flicks ever made.

I remembered reading somewhere that, in his real life, Quaid was a private pilot and a bit of an aviation enthusiast, so I invited him up to take a look at the helicopter. Without hesitation, he tossed his cigarette aside and made his way up the steps to the pad.

I spent the next ten minutes or so answering questions about the helicopter and how it compared to flying an airplane. I invited him to sit in the cockpit while we discussed the ins and outs of flying a helicopter, and he seemed genuinely engrossed in what I was telling him. I also think he appreciated that I was talking to him as a fellow pilot instead of treating him like a celebrity. He had graciously introduced himself and shaken my hand, but we'd only exchanged first names, and even though I never acknowledged that I knew who he was, I'm pretty sure he sensed that I did. I don't think there's a pilot alive who hasn't seen *The Right Stuff*.

There was one thing I knew for sure, though: If Jim came back out and found Dennis Quaid in our helicopter, he was going to slobber all over him like a dog with a new chew toy. Fortunately, we wrapped up the tour before Jim reemerged from the hospital. Even more fortuitous was the fact that Jim didn't notice Quaid, still standing about a hundred feet away, watching with interest as we prepared to leave. I knew he was sticking around to watch our departure, and because Jim was such a huge movie fan, I also knew that what I was about to do bordered on cruelty.

My conscience notwithstanding, I couldn't help myself. Hustling Jim into the copilot's seat, I expedited our run-up and lifted to a five-foot hover before keying the ICS.

"Hey, Jim. Isn't that Dennis Quaid standing over there next to the hospital entrance?"

"It sure as hell is," he responded. "Dammit! I wish I had seen him earlier."

I just chuckled to myself, made a slow pedal turn away from the hospital—then added climb power as we departed into the night.

Although it was funny at the time, I later felt sorry for what I had done and confessed to Jim that I had known Dennis Quaid was there that night; and not only that, I had even given him a tour of our helicopter. I expected him to be upset with me after learning he'd been pranked, but in true Jim Allday fashion, he laughed about it and forgave me right away.

As much as I enjoyed these brief respites from the office with Jim, I eventually reached the point where I was desperate to escape my desk job and return to the line. I was simply not well suited for the front office lifestyle, and over time, I began to feel more and more like the proverbial fish out of water. My morale had suffered, my disposition had soured, and my working relationship with Spanky Handley and Casey Ping began to deteriorate because of it.

There were definitely some philosophical differences that contributed to the tension between my bosses and me, especially when it came to personnel issues and how to motivate employees. Still, there's no denying that I could have done a better job trying to reconcile those differences had I really been committed to staying in the chief pilot's office. The truth of the matter is that, although I was burned out as a manager, the real reason I wanted to be a line pilot again was because that's what I was—a pilot.

Coming Full Circle

During the summer of 1996, shortly after I had taken the job as chief pilot, we were in the process of conducting interviews to hire another pilot. One of the résumés that caught my attention as it came across my desk was from a Vietnam veteran who, after a long and distinguished career, was retiring from the Texas Army National Guard. I had transferred from the Navy Reserve to the Guard a couple of years earlier and was assigned to a UH-60 *Blackhawk* unit in Austin. I showed the résumé to some of my buddies, all of whom had been flying for the Guard much longer than I had, and they confirmed what I already knew. Willy Culberson was the perfect fit for STAR Flight.

A native of Smithville, Texas, Willy was almost a legend in the Texas National Guard. An African American, he had become an Army warrant officer and a helicopter pilot during the late 1960s, long before the military had instituted any sort of affirmative-action program. After distinguishing himself in Vietnam, where he did two tours during a time when helicopter pilots were lucky to survive just one, he returned stateside with two Air Medals and a Bronze Star to his credit. In 1974, he transferred to the Army National Guard, where he retired as a master army aviator with twenty-eight years of service.

Although we interviewed several other well-qualified pilots, I knew right away that Willy Culberson was our guy. Fortunately, it wasn't very hard to convince Spanky Handley that we should hire him before somebody else did, and just like that, we had ourselves a new pilot.

Now, five years after we had hired him, I went to Willy and asked him for the same favor that Mike Phillips, when he'd wanted to return to the line, had asked from me. Willy agreed to become chief pilot, and Spanky Handley and Casey Ping were quick to approve the swap.

In fact, I think they were probably just as happy as I was about the deal. I thanked Spanky and Casey for allowing me to return to the cockpit full-time, and I promised I would do my best not to cause them any more trouble.

If my memory is correct, I think I kept that promise for about a week and a half.

Casey Ping, STAR Flight's first program manager—Prior to his appointment in 1999, STAR Flight's command structure comprised three different employers.

Willy Culberson, the man who liberated me from the chief pilot's office—Willy had a superlative military career in the U.S. Army and Texas National Guard.

Stan Wedell (left) and Mike Self (right)—They kept us in the air 24/7 by repairing helicopters as quickly as we could break them (round the clock, day and night—rain or shine).

A bird's-eye view of the Austin skyline from the EC-145's sliding cargo door—Note the swimming pool in the apartment complex (white arrow), completely devoid of swimmers, . . . of course.

9

The Leper Colony Anthology

(Volume One)

One of my biggest achievements as chief pilot was convincing my bosses that we should implement a practice that we had routinely used in the Navy during overseas deployments. Known as "battle rostering," it meant that everyone was assigned to a specific crew, and the objective was to maximize efficiency and promote better coordination between crew members during high-risk operations. At STAR Flight, each battle-rostered crew consisted of three members—a pilot, a paramedic, and a nurse.

Additionally, at least one of the two medical crew members aboard every flight had to be certified as a crew chief. Critical to the success of every public-safety mission, the crew chief was the pilot's right-hand man. If the pilot was the captain of the ship, then the crew chief was certainly his second-in-command. The pilot depended on him to be his eyes behind and below the aircraft and also to help him make difficult decisions. It was a huge responsibility, and only after he had completed an extremely rigorous training regimen was a crew member deemed to be a qualified crew chief.

During firefighting missions, the crew chief was like an artillery spotter. He would let the pilot know if the water he was dropping was actually hitting the target, and if it wasn't, he would help him to make corrections on the subsequent passes. Using feedback from the crew chief, the pilot could fine-tune his drops by adjusting his altitude and airspeed in order to lay down an effective, fire-suppressing spray pattern.

During rescue operations, the crew chief was even more indispensable. Picking someone up on a hoist could be a risky proposition. Because the business end of the cable was located directly under the helicopter, it was impossible for the pilot to see what was going on from the cockpit. Instead, he had to rely on the crew chief's verbal commands to position the aircraft properly and to help him maintain a steady hover—often for an extended period of time. This required a tremendous amount of trust and coordination between the crew chief and the pilot, but when both were perfectly in sync with one another, it was like a well-rehearsed ballet.

Needless to say, because the pilot relied on him so heavily, a sub-par crew chief could make a pilot's job that much harder during some of the more complex public-safety operations. An *exceptional* crew chief, on the other hand, could lessen the pilot's stress level during the most challenging of missions and could even make a mediocre pilot look good at times.

Now that I was about to become an everyday pilot again, I had an important decision to make. Even though battle rostering was generally a good thing, there's no denying that it was an even better thing if you happened to be rostered with a good, solid team of professionals who knew how to perform their duties well—especially when their ability to perform those duties well was what brought you back in one piece. With that in mind, I made it my top priority to enlist the services of a good crew chief when I returned to the line in the fall of 2000.

As chief pilot, I had been fortunate enough to observe each of the STAR Flight crew chiefs during various training flights, and there wasn't a single one of them with whom I would have hesitated to fly on any given day. That said, when it came time to pick one, there was no contest. I already knew who I wanted.

There was just one problem. The candidate I had in mind was someone who, because he was nearly as opinionated as I was, had logged almost as much time in management's doghouse as I had—and in order for this plan to work, I needed to persuade our bosses to let the two of us fly together as a team. I hoped I could convince them that, by assigning two firebrands to the same crew, it would make it easier for them to keep a watchful eye on us.

This was the genesis of what would eventually become known in STAR Flight lore as the "Leper Colony," the same name given to a B-17 crew in the classic Hollywood film *Twelve O'clock High*. It was a crew made up entirely of misfits.

Introducing the Leper Colony Players

The Renegade

Not long into my career at STAR Flight, during one of my shifts at Brackenridge Hospital in late 1993, I happened to be flying with a nurse by the name of Lourdes Maier. Lourdes had served on the interview committee during my hiring process a year and a half earlier, so I guess you could say she was partially responsible for the life I was enjoying as a STAR Flight pilot. She had spoken several times about her son, but I had never met him.

On this particular day, we had just returned from a flight, and Lourdes, along with the medic who was flying with us that day, was still in the ER with our patient. I finished refueling the helicopter and headed downstairs to finish my paperwork. When I opened the door to the crew quarters, I was surprised to see a skinny young kid making himself at home—sprawled on our couch, eating our food, and watching television. Sporting a semi-spiked hairdo (emblazoned with purple streaks) and an earring, he looked to be in his late teens or early twenties.

"Can I help you?" I asked him.

"Nope," he replied as he lay there looking back at me with a couldn't-care-less look on his face. "Just waiting for my mom."

"Do you belong to Lourdes?" I asked him.

"Nope—but she *is* my mother," he answered sarcastically.

"Well, go right ahead and make yourself at home," I responded with an equal dose of sarcasm.

I didn't hide the fact that I was somewhat annoyed by his royalist occupation of our living space. I think he also sensed that I was less than impressed by his personal appearance, so from that point forward, there wasn't a whole lot of conversation between us. I left him sitting on the couch and retired to the office to take care of my administrative duties.

Wow! I thought to myself. *Lourdes raised this kid? I wonder what went wrong.*

I didn't think a whole lot about my encounter with Lourdes Maier's son after that day. A few years later, in 1996, Lourdes left STAR Flight, reducing the likelihood that Stef Maier and I would ever cross paths again. At the time, I had no way of knowing that the young reprobate I had found lounging on my couch that day would someday ask me to be a groomsman in his wedding—and that I would be honored to do it. I certainly didn't think he would someday earn my trust to the point that I would be willing to bet my personal welfare on his judgment and expertise.

On October 3, 1999, as Stef Maier was in the process of becoming a medic for STAR Flight, I was the pilot on his first training flight. Over the next couple of years, I watched him develop into the best crew chief I had ever seen. Stef wasn't just the best crew chief at STAR Flight. I would stack him up against any of the highly trained crew chiefs with whom I had flown in the Navy.

By the time I returned to the line and began putting my crew together, my perception of Lourdes Maier's son had changed dramatically. Oh, he was still petulant. But the counterculture exterior he'd displayed on the day I had formed my initial opinion of him belied the character and dedication that had since manifested in him as a young crew chief at STAR Flight.

I suspect that Stef's perception of me had changed slightly as well. He must have pegged me as a straitlaced curmudgeon the day I looked him over indignantly in the crew quarters—but after watching me mix it up with my bosses every now and then, I think he realized the two of us were more alike than either party would have imagined that first day. At any rate, we were about to spend a lot of time flying with each other over the next decade or so, and in the process, we wound up becoming good friends. Ironically, especially in light of our respective initial opinions of one another, it was a friendship based on mutual trust and respect.

After Stef and I began flying together, it wasn't long before our camaraderie extended beyond the work place. We both had an affinity for Tex-Mex cuisine and Shiner Premium beer, and located within walking distance from our crew quarters was the one place to go when you wanted to wash down some good chili-laden enchiladas with Kosmos Spoetzl's golden lager. So, in addition to logging tons of hours together in the air, Stef Maier and yours truly, as the original members of STAR Flight's Leper Colony, could also regularly be found unwinding at one of the best hangouts in Austin.

Zoob and Pepe

Although they weren't officially members of the Colony, there were two individuals who certainly rated status as "honorary" Lepers. The Texas Chili Parlor, my favorite haunt, was in danger of closing its doors around the time I stepped down as chief pilot and returned to the line. It seemed the owners had gotten into trouble with the Texas Alcoholic Beverage Commission, and the state comptroller eventually seized the place and slapped a tax lien on it. Scott Zublin, or "Zoob" as he was known to regular Chili Parlor patrons, stepped in and bought the iconic bar and restaurant just in time to save it from the auctioneer's gavel, making him a hero to those of us who felt an emotional attachment to the place.

Zoob, who made the transition to restaurateur after twenty-three years in the oil fields, routinely went out of his way to accommodate us as we transformed the Texas Chili Parlor into the Leper Colony's version of a military officers' club. The same can also be said of Jose "Pepe" Lozano. A fixture behind the Parlor's historic bar, Pepe was one of the most dedicated Chicago Cubs fans on the planet (certainly in Austin). Yet any time I walked in while his beloved Cubs were playing on television, without my even asking him to do it, he would magnanimously switch to the channel broadcasting the Texas Rangers' game. In addition to his duties as chief mixologist, he also had the unenviable task of providing adult supervision to my crew and me, not to mention the rowdy cadre of hooligans we frequently brought along with us. It would be difficult to tell the story of the Leper Colony without including Zoob and Pepe in the cast of characters. Every eighth day, for the better part of a decade, we gathered at the Chili Parlor for something we referred to as "transition night."

At that time, the STAR Flight crew schedule had evolved into a two-days-on, two-nights-on, four-days-off model. The first two days, we worked from seven o'clock in the morning to seven o'clock that evening. Then we would go away for twenty-four hours before returning for two consecutive night shifts. After that, we were off duty for three and a half days, at which point we returned to start the cycle again.

Transition night was the term we used to describe the ritual that took place at the end of that second day shift. The stated objective was to head over to the Chili Parlor and stay up as late as we could. Oddly enough, "as late as we could" usually turned out to be 2:00 a.m., closing time. Then, in an effort to adjust our circadian rhythms for the two night shifts that loomed ahead of us, we would sleep in as late as we could the next morning.

This was the routine for close to a decade. Every transition night, we would leave the crew quarters, unzip the tops of our flight suits, and then roll them down far enough to tie the empty sleeves around our waists. Next, over our tee-shirts, we donned Hawaiian shirts, the official Leper Colony dress uniform. In addition to making a colorful and lively fashion statement, this helped to hide the otherwise conspicuous sleeves from our rolled-down flight suits.

Once properly attired, we would rendezvous at the Chili Parlor, where, in addition to watching baseball, we spent the remainder of the night reviewing our performance on recent missions, discussing geopolitics, and coming up with ways to make the world a better place for all mankind. Occasionally (if I'd had enough to drink), I would enrapture my colleagues, as well as everyone else in the bar, with sea stories from my days as a naval aviator.

During one such monologue, I recounted how we had ceremoniously performed "carrier quals" (aircraft carrier landing qualifications) at the Cubi Point Officers' Club in the Philippines. It was a rite of passage, and it involved running at break-neck speed and fearlessly launching yourself, chest-first, onto the bar with enough momentum to slide to the other end, where two of your squadron mates waited on either side with a belt, which was held tautly between them—just high enough for you to slide beneath it.

In order for your trap (arrested landing) to be successful, you had to raise your feet and catch the arresting cable (the belt your buddies were holding) with the back of your legs before you flew off the far end of the bar and smashed your face on the floor.

Enter the third member of the Leper Colony.

El Dangeroso

The Texas Chili Parlor first opened in 1976. To my knowledge, the only person who's ever been thrown out of the place since that time was my flight nurse—James Richard, aka J.R., aka "Junior" Esquivel.

"Junior," as I preferred to call him, was easily the best choice to fill the flight-nurse/rescuer billet in the Leper Colony. When Stef Maier and I set about trying to recruit a flight nurse, we were looking for someone solid, both in spirit and body, to fill the role as rescuer in our crew. If J.R. Esquivel was anything, he was rock solid. Even though he was in his mid-thirties (slightly older than most of the other medical crew members), I would have picked J.R. over anyone else at STAR Flight to be my partner in a bar fight. In fact, I think he once *was* my partner in a bar fight, but that's another story.

J.R had moved to STAR Flight from another air ambulance program at about the same time we were putting the Leper Colony together. He came to STAR Flight, in large part, because he enjoyed the physical challenges associated with the various public-safety missions we routinely flew, especially rescues. A fellow University of Texas alumnus, he shared a lot of common interests with yours truly, including an avid interest in the Longhorn athletic teams. This worked out well because, even though he wasn't a UT graduate, Stef Maier was also a passionate 'Horns fan. In short, J.R. was a seamless fit for the Leper Colony, which was why Stef and I both agreed he should be our flight nurse.

Nothing better illustrates just how well J.R. Esquivel assimilated into the Leper Colony than the story of one of our more memorable nights at the Chili Parlor. After listening to me tell that story about how my fellow naval aviators and I had conducted carrier quals at the Cubi Point O'Club, J.R. decided this sounded like great fun. It was just about closing time, and my crew and I were the only people left in the place, along with Zoob and Pepe. In fact, ours was the only table that didn't have chairs stacked on top of it.

Pepe was in the process of shutting down the bar, and Zoob looked like he was just about to show us to the door, when J.R., out of the blue, asked if it would be all right to use the Chili Parlor bar for *his* carrier quals.

Zoob just shook his head and said, "I don't think that's a good idea, Junior (Zoob called him 'Junior' as well)."

"Why not?" J.R. persisted.

"Well, for one thing," Zoob replied, "the furniture in this place is thirty years old, and there's a lot of irreplaceable memorabilia around that bar over there. I don't want to see any of it destroyed while you're trying to fly like Superman."

Stef and I tried to convince J.R. it was time to call it a night, but he had made up his mind. Instead of leaving, he reached into his pocket, took out his wallet, and produced ten brand new, fresh-from-the-ATM twenty-dollar bills.

"How about now?" J.R. asked, waving the pristine cash at Zoob.

Without hesitating, Zoob took the two hundred dollars from J.R.'s hands, whereupon he turned to Pepe, waved his arm in a wide sweeping motion, and yelled, "Clear the bar!"

Somewhat surprised by Zoob's mercenary acquiescence to J.R.'s request, Stef and I dutifully stationed ourselves at the end of the bar and waited. Because we were in flight suits, neither of us was wearing a belt, so Pepe graciously loaned us his.

J.R. ceremoniously removed his Hawaiian shirt, slipped his arms, one at a time, back into his flight suit, and then slowly zipped it all the way to the collar. Standing almost at attention, he very deliberately finished his drink as if he were about to depart on a kamikaze mission, and then he gently placed the empty glass back on the table. In a display of precision that temporarily concealed his advanced state of inebriation, he smartly marched to the end of a long hallway that led back to the kitchen. Hidden from our view, he prepared to start his run.

Because of the layout in the Chili Parlor, a straight-in approach to the bar wasn't possible. The hallway from which J.R. was about to emerge was perpendicular to the bar, which was going to necessitate a ninety-degree turn onto a very short final approach.

When J.R. started his run, Stef and I couldn't see him, but we could hear him coming down the hall, picking up steam. He came into sight just as he attempted to round the corner in a full-on sprint. Unfortunately, the floor was a little slick where he planted his foot to make the turn, causing J.R. to go down hard and crash into a metal cabinet that contained drawers full of tortilla chips.

Although the crash sounded catastrophic, J.R. emerged without a scratch, and only a few dollars' worth of chips were broken in the unsuccessful landing attempt. Stef and I had a good laugh and handed Pepe's belt back to him, thinking that was the end of the exercise.

Undaunted, J.R. returned to the end of the hall and steeled himself for a second attempt. This time he successfully negotiated the turn to final and launched himself—just in time to land chest-first on top of the bar instead of *head-first into* the bar. He was carrying an impressive amount of momentum, given the tight pattern he was forced to fly, and it was only after we saw him careening toward us at high speed that Stef and I realized we had prematurely returned the arresting cable to Pepe.

We watched, helplessly, as J.R. went sliding past us and off the end of the bar. This time he crashed into a wood-framed glass showcase next to the door. By some miracle, the glass remained intact, and once again, the damage was only minimal.

Shaken but uninjured (at least there were no visible signs of serious trauma), J.R. groaned slightly, . . . then rolled over on his back and looked up at Stef and me.

"Let's go home," was all he said.

As exciting as J.R.'s carrier quals were, they weren't what ultimately got him kicked out of the Chili Parlor. That came a few years later, on a night when Zoob was in the process of negotiating a deal with Quentin Tarantino, the Hollywood film director, to shoot part of the movie *Death Proof* in his bar.

It happened to be our transition night, and the Leper Colony was holding court next to the table where Tarantino, along with Kurt Russell and Goldie Hawn, was having dinner and some drinks with Zoob. Without warning, for reasons that are still unclear to this day, J.R. decided to start heckling Quentin Tarantino.

One thing led to another as the evening wore on, and J.R. continued to make a nuisance of himself, incessantly harassing the now-agitated director. Finally, as J.R. was leaning back in his chair to berate Tarantino one last time, the chair went all the way over, and J.R. wound up on the floor, looking straight up at Goldie Hawn. She looked down at J.R. with a look of concerned amusement and asked him if he was all right.

Without waiting for J.R to answer her question, Zoob, who was seated on the opposite side of the table, pushed his chair back and stood up with a fair amount of purpose. He walked around to where my *slightly* inebriated flight nurse was now lying horizontally in his chair, much like an astronaut waiting for liftoff. Then he looked down at J.R. and, with a stern expression on his face, informed him (in no uncertain terms) that his night was over.

You couldn't really blame Zoob. He and Pepe put up with a lot from the Leper Colony. It was not unusual for us to close the Chili Parlor down on transition nights, often after we had undoubtedly run off some of the more timid patrons who weren't accustomed to such raucous behavior from three guys ridiculously dressed in flight-suit/aloha-shirt ensembles. Stef always claimed that we were able to maintain our good standing because the proceeds from our weekly gatherings helped to put several of the Parlor waitresses through college, but because I did the math, I happen to know this wasn't true.

Thanks to some high-speed, low-drag accounting software on my home computer, I was able to determine that during the years the Leper Colony flew as a crew, I personally dropped a total of $13,320.03 at the Texas Chili Parlor. Assuming my fellow Lepers spent equal amounts, that's $39,960.09. We were usually accompanied by several guests, so let's call it $50,000 even. That was just the tab, of course, but from that number we can calculate (fairly accurately) the cash tips we left for the waitresses. Depending on how many drinks we'd had and how many times Pepe had been forced to apologize to the other patrons for our unruly behavior, we usually tipped anywhere from 20 to 30 percent. So, at the very most, we had left a total of $15,000 in tips, and fifteen grand wouldn't even pay for one waitress to attend one semester.

No—looking back on it, I think it's more likely that Zoob and Pepe tolerated us because of Rose Marie Caputo.

Rose Caputo was a registered nurse, and in the predawn darkness of a foggy November morning in 2004, she unwittingly drove her car into a flooded low-water crossing while on her way to work. Even though it was located on a major thoroughfare, the flooded crossing had inexplicably been left unbarricaded, and Rose, while travelling at high speed, plunged into the water and was trapped inside her car.

Suddenly, Rose, who was on her way to start a shift at the Heart Hospital of Austin, where her job was to provide for *other* people in life-and-death scenarios, found herself about to be immersed in the rapidly rising torrent—and at a time when the weather was barely flyable. Of course, the fact that the weather was lousy on a morning when someone needed to be rescued was not unusual. People rarely needed to be pulled from floodwaters when the weather was good and the visibility was clear.

As fate would have it, it turned out that Rose Caputo's accident had occurred during the waning hours of a Leper Colony night shift.

Onward through the Fog

November 22, 2004

I was the only one awake when the pager went off a little after 4:00 a.m. that morning. This wasn't unusual, as I was rarely able to sleep during night shifts. I was always fearful we would be dispatched just as I was slipping into a deep, restful sleep. Even dating back to my time in the Navy, the prospect of launching into the night while only half-awake had never really appealed to me. I know it seems counterintuitive, but I logged most of my napping hours on day shifts. I knew that as soon as I opened the door to the crew quarters, the daylight would quickly resuscitate me from whatever level of slumber I'd managed to achieve just before the pager went off.

Stef Maier and J.R. Esquivel, on the other hand, were never afflicted with the fear of sleeping on night shifts, and even though they were quick to the helicopter, my head start usually gave me enough time to check the weather one last time and still beat Stef and J.R. up to the pad. It was no different this time. As soon as the pager sounded, I was at the computer, checking the latest observations.

The weather that night was not good at all. Because the call was dispatched as a water rescue, it was considered a public-safety flight, which meant we weren't bound by the usual FAA weather minimums in making our decision on whether to launch or cancel the mission. That notwithstanding, if I was going to hover the helicopter for a rescue, I still needed some sort of visual reference. When I got to the top of the stairs, I could barely see to the far side of the helipad. It had rained heavily just a few hours earlier, and there was dense ground fog everywhere.

My first inclination was to cancel the flight for weather, but I noticed that if I looked straight up, I could just make out the soupy outline of an almost-full moon. From the information that had come across the printer when we were dispatched, we knew there were already first responders on the scene, and they had confirmed there was a victim in the water who needed our help. This was the real deal.

When Stef and J.R. made it up the stairs, I told them I felt comfortable punching up through the thick ground layer. That way, we could at least fly to the scene before making a decision whether to abort the mission or attempt the rescue.

Because I was the pilot, they usually left the go, no-go decisions to me, but I still liked to give them a chance to pull the plug on flights in bad weather. The thing is, in all the years we flew together, I never remember them backing off from a mission if I said I thought we could make it. They were not reckless by any means, but saying no, if there was the slightest chance of success, wasn't in their DNA. I think they looked to me to make good decisions with regard to their safety, and it was a responsibility I didn't take lightly.

It was an eerie sight when we climbed out on top of the fog and into a beautiful, clear night sky. As we headed south to the flooded low-water crossing that lay just across the Travis County line, the moonlight brilliantly reflected off the top of the fog layer beneath us. Shining from inside the fog, the hue from some of the brighter lights on the ground was barely visible. This gave us hope that, using the overhead lights from the first responders' vehicles, we would be able to pinpoint the low-water crossing where the victim was stranded.

Sure enough, when we were just inside a mile from our assigned GPS coordinates, we began to see multiple sets of red and blue lights faintly flashing inside the still-heavy fog layer. The good news was that we had managed to find the scene. The bad news was that, even when we flew directly over the top of it, all we could see were the colorful eruptions of light in the ground-based clouds beneath us.

"Well, what do we do now?" Stef asked from the copilot seat.

In 2004, we were flying EC-135 helicopters, and they weren't equipped with rescue hoists. Neither did we have night-vision goggles back then (not that they would have helped in the fog).

Because we didn't have a hoist, we would land after arriving on the scene of a rescue and rig something called a "short-haul" line. The crew chief would secure one end of a rope to the bottom of the helicopter. Then, while we were still on the ground, the rescuer (who was tethered to the other end of the rope) would station himself where the pilot could see him, directly in front of the aircraft.

After receiving a thumbs-up from the rescuer, the pilot would lift off, then climb and slide forward until the crew chief advised him that the aircraft was directly over the rescuer. Seated on the cabin floor with his feet on the starboard skid, the crew chief was looking straight down at the rescuer from his vantage point behind the pilot. After receiving one last thumbs-up from the rescuer, the crew chief would command the pilot to slowly climb straight up, putting tension on the short-haul line.

From there, it was a matter of lifting the rescuer into the air and delivering him to the victim, a task that required precise communication between crew chief and pilot. Giving him commands over the ICS, the crew chief had to vector the pilot, who couldn't see the short-haul beneath him, into position for the rescue. This was not an easy thing to do. The crew chief was, in essence, flying the aircraft verbally from his perch on the skid.

The first order of business, if we were going to attempt a rescue in the fog, was to figure out how to safely get the aircraft on the ground so we could deploy J.R. on the short-haul line. The on-scene commander, who was talking to us by radio from the ground, said there was a field big enough for an LZ about a quarter mile from the victim. He also said the woman who had been trapped in the low-water crossing was still inside her vehicle, and the water was rising all around it.

To help us locate the LZ through the heavy fog, we asked him if he could station two or more of his vehicles so that the beams from their headlights would intersect at the center of the LZ. He said he would do his best to accommodate us, and within a matter of minutes, we had several sets of headlights and a couple of spotlights pointed at the middle of the field from three sides. Even though we couldn't actually see the ground through the fog, this gave us a pretty good indication where the LZ was. We asked him if there were any overhead obstructions around the LZ, and he said there were none that he could see, but then he added that he couldn't "make any promises" because of the fog.

I told Stef and J.R. that, once we were on short final, we might be able to move enough fog with our rotor wash to get into the LZ and land. They both agreed that, given the gravity of the victim's situation, we should at least give it a try. The first responders on the ground had no way to reach the woman trapped in her car, so even though nobody would come right out and say it—if we couldn't get our short-haul to her, the woman was most likely going to drown.

We set up for our approach, and because of the poor visibility, we came in even steeper and slower than we normally would. If my fog-moving theory turned out to be wrong, I wanted to be able to wave off the approach before we actually committed ourselves to the LZ.

As we came down to within about 100 feet of the fog layer, the glare from our night sun became too bright, and I quickly turned it off. This allowed me to see the intersecting light beams more clearly, and the orange afterglow from the still-hot night sun provided just enough light for me to judge how far we were from the top of the layer. I continued the slow, meticulous approach, and the radar altimeter indicated we were a little more than 300 feet above the ground when the vortices from our main rotor blades began to stir the thick fog, now only 50 feet or so below us.

"I think this is going to work," Stef said over the ICS.

Reminiscent of the movie scene in which Moses parts the Red Sea in *The Ten Commandments*, the fog began to roll back farther and farther as we descended into the LZ. A few seconds later, I could see the ground. There was no time to relight the night sun, but the headlights and spotlights from the emergency vehicles were more than adequate at this point. We touched down softly in the wet grass, and Stef and J.R. wasted no time bailing out to begin rigging the short-haul.

While they were busy with that, I raised the on-scene commander on the radio. I told him I needed his people to repeat the targeting operation—just like they'd done to guide us into the LZ—only this time, the woman we were trying to rescue was to be the focal point.

He asked me if he should wait until after we departed with J.R. on the short-haul to move his vehicles, and I told him to "go ahead and start repositioning now."

I had come straight down through the fog to get *into* this LZ, and I planned on getting out the same way, by going straight back up. I didn't need any external lighting to accomplish that. All I needed was a good crew chief and a steady hand. I knew I could count on Stef to satisfy the first requirement—the second was up to me.

As the emergency vehicles were leaving us to station themselves around the low-water crossing, Stef climbed into the cabin and reconnected his ICS cord.

"I doubled the line down to forty feet," he said.

This was critical because we were depending on our rotor wash, just as it had during our descent into the LZ, to displace the fog beneath us during the rescue. The short-haul line was eighty feet in length, and that would have left J.R. hanging too far below the aircraft. We would have had to dip him into the fog before we were close enough to it for the rotor wash to be effective. I had actually intended to tell Stef to shorten the line, but in my haste to tell the on-scene commander to reposition his vehicles, I had forgotten to let him know. Fortunately, he thought of it on his own, which demonstrated why I had picked him to be my crew chief in the first place.

After receiving a thumbs-up from J.R., I raised the collective, and Stef began giving me commands to "continue up and slide forward."

No longer able to see much of anything from the cockpit, I was not only depending on my crew chief to keep me directly over J.R. as we climbed, I pretty much had to trust him to keep me inside the LZ until we were high enough to clear the surrounding obstacles. It was hard to hold a steady hover in the fog, so I was extremely pleased when I finally felt the strain coming onto the short-haul line. A couple of seconds later, we gently pulled J.R. from the ground beneath the helicopter.

"Come straight up," Stef said, although I don't think he realized that coming straight up in the fog was easier said than done. I'm sure he was expecting me to continue a slow, gentle ascent, but as soon as he informed me that J.R. was off the ground, I grabbed a larger-than-normal amount of collective and expedited our departure from the LZ.

By the time we had climbed above the fog with J.R. suspended on the short-haul line, the first responders had repositioned themselves around the low-water crossing. We told the on-scene commander that we were ready to attempt the rescue, and he gave us a quick, not to mention discouraging, update on the victim.

While we had been rigging the short-haul, the force of the floodwater had carried the woman's vehicle off the pavement and into the creek, where it had then rolled about forty-five degrees to one side before coming to rest against a submerged obstacle in the middle of the swift current. As it was filling with water, she had managed to kick one of the doors open, and now, with the water rapidly rising around her, the woman was desperately clinging to the luggage rack on top of her inundated vehicle.

Needless to say, this made her situation even more precarious. Now, it was possible that the same rotor wash we were counting on to move the fog from around her during our descent to deliver the short-haul might actually cause her to lose her grip on the slippery metal rack.

This meant that we needed to be especially precise with our short-haul delivery so that J.R. could work quickly. The longer we exposed her to our rotor wash before J.R. was close enough to grab her and secure her to the short-haul line, the greater the likelihood we might actually contribute to her death in the process of trying to rescue her. Stef Maier was about to earn his pay. It was going to be up to him to direct me in quickly and accurately so that J.R. could complete the rescue before she slipped into the rushing floodwaters and disappeared into the night.

Not surprisingly, Stef was up to the task. Using the intersecting beams of light from the emergency vehicles as if they were crosshairs on a rifle scope, he guided me down through the swirling mist. When the fog finally rolled back from around the woman we were trying to rescue, we had come to a hover directly above Stef's intended target. She was almost within J.R.'s reach. Stef could see the frantic woman struggling as she clung to the luggage rack with all her strength against the raging current and our punishing rotor wash. Stef gave me one last command.

"Slide forward three feet."

That was all J.R. needed to get the rescue ring around her and grab the luggage rack himself. He pulled himself close enough to her face to make himself heard over the noise from the rushing torrent and the hovering helicopter.

"You obviously didn't panic!" J.R. yelled to the woman, who was exhausted and shivering from exposure to the frigid water.

"How do you know that!"? she yelled back.

J.R. grinned at her.

"You wouldn't still be here if you had!"

Finally able to let go of the luggage rack, the woman grabbed J.R., wrapping her arms around his neck. Then to J.R.'s surprise, she kissed him and thanked him for saving her life.

As soon as Stef saw the thumbs-up from J.R., we began lifting the two of them straight up. Normally, we would have stayed as low as possible until we were able to set them back down, but it was going to take a few minutes for the first responders to illuminate our LZ again, and because of the poor visibility, we had to be sure we weren't going to drag our dangling passengers through hidden trees or power lines. We gently flew them around in circles at several hundred feet, just above the fog, waiting for the first responders to move their vehicles.

Because they had already done it twice, it didn't take them long to target our LZ one last time. Within a matter of minutes after we had first seen her through the fog, the woman was safely on the ground—wet and cold, but uninjured.

Naturally, when we flew back to Brackenridge Hospital that morning, we knew very little about the woman we had pulled from the flooded creek. She hadn't required any medical treatment, so we'd left her there at the scene, in the care of the first responders. Because she was wearing scrubs, J.R. had assumed she was a nurse, but we didn't know much else about her.

We didn't even know her name.

15 Days Later

It was business as usual on our transition night. The Leper Colony had convened at the Texas Chili Parlor, and we were several rounds into the evening. Just like every other night we were there, we were undoubtedly making too much noise and probably annoying a few of our fellow patrons when Zoob came over to our table and pulled up a chair, turning it around so he could rest his arms on the back as he straddled the seat. He had a serious expression on his face, so, naturally, we were anticipating a lecture on our disorderly conduct.

Much to our surprise, however, he began telling us about the details surrounding a recent water rescue and asked if we knew anything about it. We were surprised to learn that the woman we had pulled from the flooded creek that foggy morning was one of Zoob's good friends. In fact, it turned out that Rose Marie Caputo was a friend to both Scott Zublin *and* Pepe Lozano.

After learning we were the STAR Flight crew who had rescued Rose, Zoob patted us on the back and generously comped our tab that evening. From that night forward, as far as Zoob and Pepe were concerned, it was almost as if the Leper Colony could do no wrong at the Chili Parlor.

Of course, that didn't stop us from giving it our best effort.

THE LEPER COLONY ANTHOLOGY (Volume One)

You gotta learn that if you're gonna take the last shot of the game, it's either gonna go in, or it's not gonna go in, and you're either gonna be the hero or the goat.

—Tom Heinsohn (NBA player and coach)

Even though I'd have to say that Rose Caputo's rescue was more dramatic than most, it wasn't the only nail-biting mission in which Stef Maier, J.R. Esquivel, and yours truly were involved. There were a couple of occasions when the rescue itself was only a sidebar to the main storyline. Even though both of the rescues to which I'm referring were ultimately successful, both times the Leper Colony's legacy had hung in the balance right up to the point where our skids touched down on the Brackenridge Hospital helipad.

During the first of these two operations, it was raining hard—I mean *really* hard. The actual rescue took place about twenty miles east of San Antonio, but that wouldn't be the most electrifying part of the night for us.

When the heavy rains started just before sundown, a rancher had gone out on his ATV (all-terrain vehicle) to check on his goat herd. The goats had been grazing in a low-lying field that was prone to flooding anytime the Guadalupe River rose out of its banks (contrary to what you're probably thinking, the fact that the livestock in this story comprises a herd of goats has nothing to do with the lead-in quote at the beginning of the subchapter). The storm intensified dramatically, and when the rancher failed to return from checking on the aforementioned goats after several hours, his family called 911.

As you might guess, the weather that night made the thirty-minute flight to the scene a little dicey. Nevertheless, we arrived over the ranch with enough fuel to search for about forty-five minutes and still make it back to Austin. There were several Guadalupe County sheriff's deputies there with the rancher's family, but they simply didn't have the resources to search a flooded area as large as this one. They told us where to begin our search, and we set about trying to find the rancher from the air.

This proved to be difficult, given the extent of the search area. Instead of looking up and down a swollen creek or river, as was usually the case in these types of rescues, we were searching dozens of acres of flooded land that was heavily wooded. We knew that if we were going to find him alive, we would most likely find the stranded rancher in a tree. We literally had to shine our night sun into every single tree as we slowly flew over it at low altitude. Part of the challenge was figuring out a systematic pattern that would allow us to be positive we were searching each and every tree without wasting time looking in trees we had already cleared.

We searched, without success, until we had reached our bingo fuel state, at which point we contacted Casey Ping, the program manager, by radio. We asked him for permission to refuel in San Antonio so we could continue the search. Because of the heavy rains, he was concerned that we might be needed for rescues back in Travis County, but he ultimately decided we should complete our search for the missing rancher before returning home. This was welcome news to the family, who relayed their appreciation to us through one of the deputies there with them.

So that we would know where to resume searching when we returned, J.R. activated several Cyalume light sticks, tied them together on several feet of parachute line, and then tossed them into the branches of the last tree we had cleared. In addition to that, I created a waypoint on our GPS receiver to mark the spot, and then we made the short flight to San Antonio International Airport. We refueled and returned in a little less than forty-five minutes from the time we had suspended the search. It had been raining hard from the time we first departed Brackenridge, but now the weather was getting even worse. The ceiling was coming down, along with the visibility, and we knew that if we were going to find our missing rancher, it would need to be soon.

When we returned to the waypoint I had created in the GPS database, the light sticks were nowhere in sight. Not knowing if the light sticks might have fallen from the tree and been carried away by the floodwaters, we decided to go ahead and pick up the search from the GPS waypoint. Just to be sure, we also made the decision, even though we might have already cleared them, to backtrack over several of the trees closest to the waypoint.

This turned out to be a prudent course of action. While lighting up a tree I was sure we had already searched (this was later confirmed by the victim), J.R. caught a fleeting glimpse of the rancher, who was frantically waving at us through the thick canopy of the huge live oak. As we flew past it, J.R. quickly tossed another ring of light sticks into the tree to mark it, and told me to come around. As we flew over him for what was now the third time, J.R. signaled to the rancher to let him know we'd seen him. Then we left him there and departed to look for a dry place to land and rig the short-haul.

This time, the light sticks were still glowing brightly when we returned with J.R. deployed on the short-haul line. The rescue itself was uneventful, save for the fact that J.R. had to convince the man to leave his coyote rifle in the tree. He had been carrying it with him at the time of the flood, and seeing as how he'd already lost his ATV, he was now understandably reluctant to leave his rifle behind as well.

Yelling over the noise from the helicopter, J.R. finally managed to cajole the rancher into the rescue ring by promising to share our GPS coordinates with him so that, when the floodwaters eventually receded, the man would likely be able to pinpoint the tree where he was abandoning his firearm. Having successfully negotiated a quick settlement to the rifle dispute, J.R. passed the good-to-go signal to Stef, and we moved the rancher to the same dry ground where we had deployed J.R. on the short-haul minutes earlier. Then we landed close by and transferred the rancher (along with J.R.) into the helicopter before returning him to his anxiously waiting family.

The decision to refuel instead of aborting the search had been fortuitous for our rifle-toting rancher, but the delay had also meant that our return flight to Austin was going to be that much more challenging. It was raining even more heavily and steadily now, and the visibility was down to less than a mile. The prudent course of action would have been to land in San Antonio and wait it out, but that would have also meant that Travis County would be without its primary rescue asset during the height of the storm. I was still relatively young and bulletproof at that time, so instead of making what would have undoubtedly been the wiser choice, I decided to fly home using a method often referred to by helicopter pilots as "flying the concrete compass."

Flying at an altitude low enough to maintain visual contact with the headlights on the highway, we headed west, staying between the eastbound and westbound lanes of Interstate 10 until we were near San Antonio. Then we flew north, using the same technique to navigate our way up Interstate 35, toward Austin. This was a slow and nerve-racking process, and had it not been for the fact that I knew there were no antennas within the confines of the highway median (I'd flown that route hundreds of times), I would have given up early on. As a last resort, I knew we had plenty of fuel on board to get an emergency IFR clearance and shoot an instrument approach if it became necessary. Much to my disappointment, after we had made it about half way home, I was suddenly forced to exercise that "last resort" option.

Quicker than you can say "little white room," the visibility went to almost zero. I grabbed an armful of collective and started climbing at about 1,000 feet per minute. We had been creeping along at around 40 knots, so as soon as I was sure we were at a safe altitude, I lowered the nose slightly on the attitude indicator—as we continued to climb—and accelerated to 80 knots, which is a much more comfortable airspeed when you're flying in the clouds.

I called Austin Approach Control to let them know we were "inadvertent IMC," and they quickly came back with an IFR clearance to Austin-Bergstrom International. I asked them if we could amend the destination to the San Marcos Airport, which was about thirty miles south of Austin and closer to our position. They said they didn't recommend an approach to San Marcos because the radar indicated a large cell in that area.

Holy *fulminations*, Batman! They weren't kidding. Just a few minutes later, the cockpit lit up like a welder's arc. This was not the same "friendly goo" I had accidentally encountered while flying the Bell 412 ten years ago. The clouds into which I had stumbled on this night contained embedded thunderstorms—and the Leper Colony had just found one of them. We were getting tossed around inside the aircraft like rag dolls, so shooting an ILS approach to San Marcos was out of the question. It was all I could do just to keep us right-side up in the turbulence.

Unlike the Bell 412, the EC-135 wasn't rated for IFR operations. The good news was, the only two pieces of equipment it lacked for such a certification were an autopilot and a standby attitude indicator. I had logged enough instrument hours not to be concerned about the autopilot, but the lack of a backup to our only attitude indicator was a different matter. The attitude gyro in this particular aircraft had a history of rolling over and dying occasionally. We had never been able to isolate the cause, and because we were a VFR-only program, it wasn't a high priority.

Now, all of a sudden, the thought of a failed attitude indicator loomed large in the back of my mind. If that thing went belly-up on me during this tilt-a-whirl ride, we were going to be in serious trouble.

Fortunately, after what seemed an eternity, we punched through the other side of the cell, and even though we were still flying IFR, the goo was a lot friendlier now. Stef talked me through the ILS to Bergstrom, just as Tom Bryan had done on the night— a decade earlier—when Tom and I were forced to shoot an instrument approach to the old Robert Mueller Airport after inadvertently flying into the fog. On that occasion, the ceiling had been so low that we were forced to spend the night at the airport. On this night, after breaking out of the clouds over the airport, Stef, J.R., and I were able to make it back to Brackenridge Hospital.

My decision to fly back that night instead of waiting out the storm in San Antonio had proven to be a boneheaded one. I was reminded of something I'd heard Chuck Yeager say one time during an interview. When asked what it was that made him such a good pilot, he responded that he had been fortunate enough to survive all of his mistakes. According to Chuck Yeager, every time a pilot scares himself without dying in the process, he becomes a better pilot.

I ended up going inadvertent IMC one more time in my career, so I was obviously a slower learner than Chuck Yeager would have wanted me to be.

For helicopter pilots, there can be only one quicker path to "goat" status than flying into a thunderstorm, and that would be running out of fuel—which brings me to that other mission that easily could have resulted in a premature and ignominious end to the Leper Colony. This one involved the high-angle rescue of someone who had fallen onto a ledge near the top of a 400-foot peak.

Packsaddle Mountain is a popular hang gliding venue, about seventy miles northwest of Austin. Instead of rescuing a hang glider, however, we were dispatched there, just after sunset one summer night (have you noticed that these rescues always take place at night), to short-haul a fallen rock climber. The ledge onto which he had fallen was totally inaccessible from the summit just above it. There was an outcropping of rock between the summit and the ledge, making it impossible for the first responders to rappel down to where the victim was helplessly marooned with two broken legs.

Allowing for the forecast winds that night, our standard fuel load would have allowed us to make the flight to Packsaddle Mountain, spend twenty minutes effecting the rescue (which would have normally been more than enough time for a short-haul), and then return to Brackenridge Hospital with a twenty-minute reserve in the tanks. There were a few STAR Flight pilots who liked to add additional fuel for missions like these, but I rarely did. In addition to the minutes it added to our response time, the problem with putting on extra fuel was that these types of rescue calls were often cancelled even before we launched.

This was just the fluid nature of the rescue business. It was not unusual for first responders from surrounding counties to call for STAR Flight, only to determine shortly thereafter that they didn't need us. Sometimes, the rescue turned out to be simple enough that a helicopter wasn't required, and other times, the patient was DOS by the time they arrived. More than once, pilots had just finished pumping several hundred additional pounds of fuel, only to be informed that the call had been cancelled. Unfortunately, this meant that, because not enough of the extra fuel would be burned off during the short flight to an in-county scene, the helicopter was now too heavy to transport patients inside Travis County.

Because I'd calculated we already had enough fuel for twenty minutes of rescue operations, I chose not to add fuel that night. This decision would have been a good one had it not been for the unforeseen complexity of the rescue—not to mention the bogus winds-aloft forecast I received from the National Weather Service.

Only after we'd arrived over the scene did we learn of the outcropping that shielded the ledge on which the fallen rock climber had come to rest. It was only about 20 feet above him, and in fact, he had fallen trying to negotiate it during his ascent to the summit just prior to nightfall. This was going to make it impossible for us to deliver J.R. on the short-haul line in a conventional manner. After some discussion on the matter, we came up with a plan to move in beneath the outcropping and put J.R. onto the ledge using a maneuver known as a "one-skid."

There was just enough room to place one of our skids on the rocks next to the victim and hold the hover long enough for J.R. to step out of the aircraft and onto the ledge. This would also necessitate a subsequent one-skid to deliver the rigged short-haul line to J.R. once he had completed packaging the patient on a backboard. Then we would have to carefully slide out from under the outcropping and up from the ledge before putting tension on the short-haul. We knew this would create a swinging motion as J.R. and the patient came off the ledge, so we would have to continue sliding away from the cliff face to avoid slamming them back into the rocks on the return swing.

Following a short and lively deliberation, the three of us decided the plan was viable, and even though it involved an elevated (pun *intended*) level of risk, we all agreed it was the best option available if the injured rock climber was going to make it off the side of the mountain that night.

Normally, I would have preferred to execute the one-skid hover with the ledge on my side of the aircraft so I could see the skid as I placed it on the rocks. I was also keenly interested in how close our rotor disc was going to come to the rocky cliff face directly above the fallen climber's resting place, but this would have required me to put the tail into what had become an exceptionally gusty wind. Instead, I decided to hover into the wind and trust my crew chief to guide me in.

Had I been flying with anyone other than Stef Maier, this would not have been an easy decision to make. Just as he had proven himself during Rose Caputo's rescue, the night he'd managed to steer me straight to her car through the fog, Stef was once again about to be called upon to confirm my faith in his crew-chiefing abilities.

Demonstrating he was up to the task, my unwavering crew chief talked me in next to the injured climber, and just as we had planned—with the tips of our rotor blades just a few feet from the side of the mountain—we deposited J.R. onto the ledge. Stef then passed the backboard and medical kit to J.R. before he and I shoved off to wait for J.R.'s assessment over the radio.

J.R. let us know right away that it was going to take more than just a few minutes to stabilize and package the patient, so I landed and shut the aircraft down in an effort to conserve our fuel. At that point, we still had enough in the tanks to complete the rescue and make it back to Austin with a twenty-minute reserve. We told J.R. to give us a couple of minutes' lead time when he was about to finish packaging the patient for the short-haul. That way we could restart the helicopter and be airborne again when he was ready to go.

Stef got out of the aircraft to rig the short-haul, and I got out just to stretch my legs a little. After about ten minutes or so, J.R. called us on the radio to let us know he was almost ready. Stef and I mounted up again, and shortly thereafter, we were delicately nestling our port-side skid against the ledge for a second time. Stef tossed the line to J.R., who wasted little time attaching it to himself and the patient.

As soon as J.R. signaled that he was ready to go, Stef told me to start sliding away from the ledge. Once we were clear of the outcropping above us, we began a slow climb. Just as we had anticipated, we induced a fairly significant swing as J.R. and the patient, who was secured to the backboard (immediately in front of J.R.), came off the ledge. I could feel our center of gravity oscillating from side to side as we descended to our LZ, nearly 500 feet below.

The short-haul was still swinging as we prepared to set J.R. and the patient down on the ground. Stef gave me a couple of well-timed sliding commands to nullify the oscillations, and as soon as the load had steadied, we gently lowered the two of them onto the ground. As soon as they were down, we quickly descended another 10 feet or so to give J.R. the slack he needed to disconnect himself and the patient from the short-haul line. Once they were detached, we moved a safe distance away and landed. Stef continued to give me commands all the way down to make sure we didn't put one of our skids on top of the free-hanging rope as we landed.

Once we were down, I debated whether or not I should shut down again to save fuel, but J.R. indicated he was ready to load the patient, so I kept the rotors turning. Stef jumped out to secure the short-haul while J.R. and two of the first responders loaded the patient aboard the helicopter. Several minutes later, we were airborne and on our way back to Brackenridge Hospital.

It was then that I noticed something was wrong. We had fought a 25-knot headwind the entire way from Austin to Packsaddle Mountain, so naturally, I had anticipated a 25-knot tailwind on the return flight. Instead, I was unexpectedly crabbing nearly thirty degrees to the right of my intended course just to maintain a straight line toward home, and more importantly, the GPS indicated our groundspeed was nearly 20 knots slower than it should have been.

This was not good! The winds aloft had shifted from west to south, and our course back to Austin was 115 degrees. This meant we weren't going to have that tailwind I'd expected. In fact, we were bucking a slight headwind component.

There was no place to get fuel along our route, and according to my calculations, this was going to be considerably tighter than I had anticipated when I'd elected not to take on additional fuel before launching from Brackenridge. Now, as I considered this unexpected turn of events, I began to grow somewhat apprehensive (to put it mildly) about our situation. I wasn't going to let myself run out of fuel and auger in, of course, but if we were forced to land short of Austin with a patient on board, it was going to reflect poorly on the Leper Colony—and on me in particular. Our patient's injuries were not life-threatening, but this was still a huge deal. If I screwed this up, there was a good chance I could lose my job—and rightfully so.

I thought to myself, *You gotta be kidding me. How unlucky do you have to be to have a headwind on both legs of a roundtrip flight?*

I tried climbing, then descending, in an attempt to find more favorable winds, but it didn't matter. It was the same at every altitude.

Meanwhile, in the back of the helicopter, where he and Stef were tending to the injured rock climber, J.R. could hardly contain himself. He kept going on about how this was the first high-angle, one-skid rescue in the history of the program. He was convinced we were all going to receive Air Medals for our efforts, and even though his crowing was mostly tongue-in-cheek, he was right about one thing: The Leper Colony had just pulled off one of the most complex and demanding rescues ever completed by a STAR Flight crew up to that time. Unfortunately, that wasn't going to count for much if we couldn't make it back to Brackenridge with the patient we had rescued.

I watched—with a steadily increasing amount of concern—as every kilogram of fuel ticked from the gauge. Every few minutes, I recalculated the time left until the first of our two engines was due to flame out and measured it against our revised ETA. As much as I hated to rain on J.R.'s parade, I owed it to my crew to let them know we had a situation on our hands. Not only that, if it became likely that we were going to land short of the hospital, we needed to expedite transferring the patient to a ground unit. This meant we would need to get an ambulance moving toward the rendezvous point so they could arrive there ahead of us.

I told J.R. and Stef that we might need to put a hold on those Air Medals until we determined whether or not we had enough fuel to make it home. Stef was appropriately concerned, but true to his unflappable nature, J.R. continued to joke about how we could still get our Air Medals posthumously if we crashed on the way home. Needless to say, the return flight was extremely stressful, more so than the actual rescue had been.

Finally, when we were exactly ten minutes from the hospital, I calculated we still had about twenty minutes of usable fuel left onboard. I didn't want to tell Stef and J.R. just yet, mostly out of a childlike fear that I might jinx us, but at that point, I knew we were going to make it all the way to Brackenridge with fuel left in the tanks.

On short final, we came across Waterloo Park, located just across the street from our helipad. He wasn't keying the ICS, but even over the noise of the engines and rotors, I could hear J.R. behind me. He was chanting, over and over, "I think I can! I think I can!"

I couldn't help thinking there was a tiny bit of irony in the park's name. Waterloo, of course, is where an overconfident Napoleon plummeted from glory when his army suffered a well-documented and humiliating defeat in 1815. The Leper Colony, on the other hand, was going to be more fortunate than Napoleon had been. They would suffer no humiliation on this night.

The "LOW FUEL" caution light illuminated just as our skids were touching down on the Brackenridge Hospital helipad. We had providentially dodged a major bullet one more time, returning home as heroes—instead of goats.

While it's true that hindsight is always 20/20, the problems we'd had to overcome returning from these two rescue missions had arguably been of my own making. They were a stark reminder of just how demanding and unforgiving aviation, especially public-safety aviation, can be. No matter how much good you manage to accomplish after each takeoff, unless you're able to log a corresponding number of landings when you're finished, the mission wasn't a success.

According to Chuck Yeager's theory, a pilot gains a measure of wisdom every time he makes a mistake and manages to survive in spite of it. In all the years we ended up flying together as a crew, I'm pretty sure I committed enough mistakes to make Stef Maier and J.R. Esquivel believe they were flying with one of the most knowledgeable aviators in the history of powered flight.

Forgotten Flights

Not every call I flew with Stef Maier and J.R. Esquivel involved a memorable rescue. By far, the bulk of our missions were EMS calls—well over 90 percent of them. Breaking it down a little further, about 60 percent of the EMS dispatches were medical calls (someone who was sick), and the remaining 40 percent were trauma calls (someone who was injured). From the pilot's perspective, the main difference in these two types of calls was the amount of time we spent on scene after landing.

If we were responding to a medical call, I would typically shut down the aircraft while Stef and J.R. assessed and treated the patients. The idea was to stabilize the patients before transporting them to the hospital. This could take anywhere from ten minutes to over an hour—although scene times longer than twenty minutes were rare, even on calls that involved the sickest of patients.

Trauma calls were an altogether different ballgame. The scenes were frequently more chaotic in nature, and instead of shutting down, we normally hot-loaded the patients while the rotors were turning. The idea was to get in and get the patients into the aircraft as quickly as possible. Then, as the medical crew tried to stabilize them en route, it was up to me to get them to the trauma center posthaste.

These were often car wrecks, some of them horrific, and you would think that each time you saw one, it would be etched into your memory. Sadly, in spite of the fact that I was often witness to the worst and most unforgettable day in someone's life, we ran so many of these calls over the years, I simply can't recall many of them—especially the hot-loads. Because we were in and out so quickly (I didn't even get out of the aircraft), the details surrounding these calls tend to run together in my memory.

One such call occurred on March 11, 2005, as Stef, J.R., and I were looking forward to another transition night at the Texas Chili Parlor. It was around 2:00 p.m. when the call came in for yet another MVA (motor vehicle accident).

A car driven by a strapping young college baseball player, named Tyson Dever, had been hit from behind by a cement truck. Tyson, who was in a Corvette convertible, was literally run over by the truck as he was stopped, waiting to make a left-hand turn. Minutes later, we landed next to the hideously mangled wreckage that had been his car, and my crew went to work. The first responders who were on scene that day credit Stef Maier and J.R Esquivel for saving Tyson Dever's life.

As I write this book, Tyson is paralyzed from the waist down, but that's not what defines him—not by a longshot. He is a teacher, a coach, and a motivational speaker who touches lives on a daily basis. Years after his accident, Tyson contacted STAR Flight to set up a meeting with the crew that had transported him to the trauma center that day. It was only by checking the historical records that we were able to determine that Stef, J.R., and I had run the call.

When we met Tyson at the STAR Flight hangar, it was a thought-provoking experience for me. It wasn't simply that I was impressed with how this young man had refused to surrender in the face of an unimaginable tragedy. I was dumbfounded as to how it came to be that I didn't remember him.

It wasn't just Tyson Dever, either. I also couldn't remember transporting another young baseball player who had survived a terrible car wreck. A little over a decade after *his* accident, that second young baseball player, Dannon King, wound up marrying my daughter, Lee Ann.

Dannon is now a significant part of my life, and yet, just as I couldn't remember the specifics from Tyson Dever's wreck, I had to admit to my own son-in-law that I couldn't remember any of the details surrounding his accident, either. Had I become so calloused over the years that I could simply erase those kinds of scenes from my consciousness? I'm fairly certain that several hours after Tyson Dever's accident, I was sitting in the Texas Chili Parlor, enjoying a Shiner with my crew as if nothing unusual had happened.

This bothered me for several days after our meeting, until I finally figured it out. The reason my crew and I were able to enjoy transition night that evening, within hours of an accident that had just changed Tyson Dever's life forever, was precisely because—nothing *unusual* had happened. By that, I mean we witnessed tragedy on a daily basis.

Looking back, if I had allowed myself to dwell on every life-changing event I had watched take place from the cockpit of that STAR Flight helicopter, I would never have been able to do what I did for twenty years. Tyson Dever overcame the tragedy that had interrupted his life by choosing not to dwell on it. Would he expect any less from me and my crew?

By the way, Stef and J.R. didn't remember Tyson's accident, either. Although they never said so, I'm fairly certain they had chosen to forget it for the same reasons that I had. Still, I have to admit—I really can't be sure *how* they dealt with those types of calls. We never talked about it with one another.

I guess we chose not to dwell on it.

Our EC-135s (1998-2006) weren't equipped with a hoist.
Rescues were accomplished using rappel and short-haul lines.

In 2006, the EC-135s were replaced with larger EC-145s.
A year later, the EC-145s were outfitted with hoists.

Stephen "Stef" Maier (left) and James Richard "J.R." Esquivel (right), my partners in crime during the years we flew as the "Leper Colony."

The Texas Chili Parlor in downtown Austin—
This is where the Leper Colony's missions were debriefed every eight days.

Scott "Zoob" Zublin, the Parlor's proprietor. *Photo courtesy of Andrea Leptinsky/ Community Impact Newspaper*

Jose "Pepe" Lozano behind the Chili Parlor's iconic bar.

Stef Maier designed these military style "challenge coins," which were presented to those we rescued. They were also liberally distributed to our fellow Chili Parlor patrons. The front of the coin features an armadillo, the only animal known to carry leprosy. The motto on the back of the coin reads, "CAN DO, ANY TIME . . . ANY PLACE."

The mangled and charred remains of Tyson Dever's Corvette in 2005.

Tyson Dever during one of his motivational speeches (left) and with the Leper Colony (below) during his visit to STAR Flight in 2015.

Dannon King with my daughter, Lee Ann, on their wedding day in 2013, nearly fourteen years after I had transported him in the back of my STAR Flight helicopter following a deadly car accident.

10

The Leper Colony Anthology

(Volume Two)

On October 14, 2000, Stef Maier, J.R. Esquivel and I worked our first shift together. We went on to work in excess of nine hundred more as a crew. During that time, we were dispatched on 1,125 flights, 103 of which were law enforcement missions. Those 103 missions primarily consisted of long, tedious searches for bad guys, most of whom were long gone before we ever got there.

It was then, during those otherwise brutally boring law enforcement missions, that I was particularly thankful we had J.R. along to keep us amused. As we droned along in our low and slow search patterns, Stef and I could usually count on J.R. to help relieve the mind-numbing boredom.

LIFE INSIDE THE DEAD MAN'S CURVE

Junior, the Entertainer

Of the 103 law enforcement missions J.R. flew with us over the years, eleven of them happened to fall during the period between Thanksgiving and December 31. On all eleven of those, without fail, J.R. would sit on the floor at the cargo door exit, feet dangling out of the aircraft, and bellow out the lyrics to "Rudolph the Red-Nosed Reindeer" at the top of his lungs. His stamina was astounding. He would do this for as long as the search lasted, sometimes well over an hour. He didn't key the ICS while this was going on, but just as he had the night we'd nearly run out of fuel, he still managed to make himself heard over the noise of the aircraft and the wind.

Stef, who was normally riding next to me in the copilot seat, would just hang his head. Then he would look over at me with an expression that can best be described as equal parts consternation and amusement. As I have already pointed out, this was not a one-time event. Every year, on every search from Thanksgiving to New Year's Day, there was J.R., singing that same stupid song to one and all. Stef and I even tried to coerce him into singing some other song—*any* other song! But alas, our pleas fell on deaf ears. J.R.'s allegiance to the foggy-Christmas hero was unwavering.

In addition to entertaining us in song, J.R. often delighted us with his "Juniorisms" on the radio. Whereas his singing was normally confined to the holidays, his humorous antics on the radio were pretty much a year-round affair. Blessed with an innate ability to use more words than you would think are humanly possible to convey a simple message, J.R. was the undisputed champion of superfluous radio communications. He regularly took great pleasure in irritating our dispatchers with his double-talking phraseology in the course of passing along routine information during our flights. This little habit actually earned him a reprimand on more than one occasion.

It didn't stop him, though. J.R. just liked to talk on the radio. There was one night, in particular, that just about sums it up. We were looking for two lost children, ages five and nine, who had wandered off from their home at three o'clock in the morning. J.R. was communicating with the police officer in charge, and the officer was providing us with the details we needed in order to optimize our search—things like the location of the house, areas where the kids liked to play, etcetera.

When the officer had finished relaying all this information, J.R. keyed up the radio and asked him to please give us a description of the kids. There was an extended period of silence on other end of the radio, as I'm sure the police officer didn't quite know how to respond to J.R.'s request.

Stef didn't actually say anything at this point, but he looked at me with a smile that begged, *Should I jump in here, or do you want to take it?*

I smiled back at Stef and keyed the ICS.

"Junior!" I quipped. "Suppose we spot a five-year-old and a nine-year-old out here at three in the morning—and they *don't* fit the description. Are we going to ignore them and continue searching until we find the ones that *do?*"

To my amazement, and to the amazement of my young crew chief as well, the police officer came back a few minutes later with a thorough description that included everything from the pajamas the kids were wearing to the stuffed animals they might be carrying. With that, J.R. triumphantly poked his head into the cockpit from behind my seat and gave us both a look of righteous indignation.

I still don't know how the police officer kept from laughing as he was passing the gratuitous descriptions over the radio.

. . . Oh yes, we did eventually find the kids. What's more, they matched the descriptions perfectly.

The only thing that makes battle psychologically tolerable is the brotherhood among soldiers . . .

—Sebastian Junger (journalist)

During the first five years of the new millennium, I spent as much time flying with Stefan Maier and J.R. Esquivel as I did interacting with my own family. To say we were a "band of brothers" would be a bit over the top (not to mention trite), but there's no denying our bond consisted of something more than just an ordinary, garden-variety co-worker relationship. The whole "Leper Colony" shared-persona had started as a joke of sorts (I can't even remember which one of us came up with the name), but as the shifts we worked together began to multiply over the years, we developed into a cohesive unit. It was an accord that went beyond flying. We learned each other's idiosyncrasies, as well as each other's strengths and faults. And we definitely learned how to prank each other.

"If you can't trust your friends, . . . you must be in the Leper Colony."

For someone who routinely had to hang himself out of a helicopter to perform his duties, Stef Maier was deathly afraid of heights. It's likely he was that way before he became my crew chief, but it's also possible that it was the result of a call we ran one afternoon after a small plane had gone missing in a thunderstorm. We were sent out to search for it, and when we found what was left of the plane, the debris was scattered over a couple of miles—along with the bodies of the three occupants.

It was obvious that the plane had come apart in the storm, and I think it may have unsettled Stef just a bit. I say this because, on the way home, he let me know that it would be just fine with him if I chose to fly at a lower altitude.

I grinned at him and said, "Stef, if we fall from *two hundred* feet, we're going to be just as dead as if we had fallen from *five thousand* feet."

"I know," he said. "But if we fall from five thousand feet, we'll have longer to scream on the way down."

Finding it difficult to argue with his logic, I humored him *that* day. From that point on, however, I took advantage of every opportunity that came my way to exploit his phobia. If we were coming back from a long flight without a patient on board, I would sometimes climb as high as I could under the pretense of looking for favorable winds. When we would get up around 10,000 feet, I'd look over at Stef, and he'd be up in his seat—sitting on his feet.

I guess he was afraid he was going to fall through the plexiglass chin bubble.

One day, at the STAR Flight hangar, J.R. and I were playing around with something called a Genie Lift, an electro-mechanical lifting device that belonged to the maintenance technicians. Mainly used for changing light bulbs in the ceiling-mounted light fixtures, it was equipped with a one-person basket, and there were controls mounted both in the basket and on the base of the lift.

We managed to coax Stef into the basket, and before he could do anything about it, we had run him up some thirty feet above the concrete floor and disconnected the battery. Disconnecting the battery disabled both sets of controls and left the youngest, not to mention loudest, member of the Leper Colony stranded in his lofty perch. Needless to say, Stef didn't find this nearly as amusing as J.R. and I did. When he began panicking and shouting at the top of his lungs, we *almost* felt sorry for him.

"Get me down!" he screamed, like a kid throwing a tantrum.

"Calm down!" J.R. shouted up at him. "You're shaking the basket!"

The basket wasn't really shaking, but I don't think Stef was cognizant of that fact because his eyes were closed throughout the entire ordeal. We let him carry on that way for several minutes before we finally took pity on him and brought him down.

Stef was madder than a wet hen when he came off the lift, and it didn't do much to mitigate his umbrage when I smugly pointed out that there was an emergency power-out button located inside the basket. If he had only opened his eyes long enough to see it, he could have thwarted our "uplifting" practical joke by simply hitting that little red button, and the basket would have gently descended at his command.

THE LEPER COLONY ANTHOLOGY (Volume Two)

Inventing new forms of entertainment at the STAR Flight hangar was a task at which the Leper Colony routinely excelled. We were often sequestered there during periods of bad weather, such as icing or thunderstorms, when we needed to protect the aircraft from the elements. There were also times when our hospital-based crew quarters were undergoing maintenance. When this was the case, we had to pack up and move from Brackenridge to the hangar. Sometimes we were stuck there for multiple shifts, and because the hangar lacked the creature comforts we enjoyed in our living quarters, we were forced to come up with new and different ways to pass the time between flights.

These pastimes often consisted of spontaneously improvised athletic competitions. One of our favorite tests of skill involved a hydraulic hoist, which was mounted on an overhead I-beam track. The track was there so that the hoist, which was actually intended to be used for such mundane tasks as lifting the transmission from a helicopter, could be traversed back and forth between the two bays inside the hangar. If one of the bays was empty, we would attach a rescue ring to the hoist, and while one competitor used the remote control to drive the hoist back and forth across the empty bay, the other two competitors would attempt to throw footballs to each other through the ring. As the ball-throwers became more and more proficient, the person driving the hoist could take the game to the next level. As he continued moving it back and forth, he could begin moving the ring up and down as well. This greatly increased the degree of difficulty and, more importantly, raised the level of entertainment for the participants.

We also competed in more traditional sports, such as baseball. This had to be taken outside the hangar, of course, since it would have been hard to justify damaging a multimillion-dollar helicopter with a baseball, even if it was for the well-intentioned purpose of maintaining crew morale. One year, the baseball season came to an unfortunate and premature conclusion when I managed to tear the medial-collateral ligament in my left knee while trying to stretch a single into a double.

As you may have figured out by now, the Leper Colony often incurred all sorts of risks that had nothing to do with actually flying missions. Contrary to what you might expect, especially considering the types of missions we routinely flew, we were actually more likely to get ourselves into trouble during the downtime between flights. Understandably, I think our bosses worried more when we were on the ground than they did when we were in the air.

This was true whether we were at the hangar or at Brackenridge Hospital, in our regular crew quarters. There were more toys available to us at the hangar, of course, but the STAR Flight administrative offices were located there as well. This made it easier for our bosses to keep us in check when we were at the hangar. When we were at Brackenridge, on the other hand, we were pretty much left to our own devices, with no supervision whatsoever. This lack of direction sometimes led to surprising and exhilarating adventures—some of which I'm actually at liberty to discuss.

THE LEPER COLONY ANTHOLOGY (Volume Two)

The One-Armed Coke Machine Bandit

For several months, early in 2001, STAR Flight fell victim to a criminal mastermind, a serial burglar whose diabolical genius rivaled that of a James Bond villain. There was a Coke machine just outside the door to our crew quarters, the funds from which went to a nonprofit crew member fund that had been set up to pay for professional development courses and things of that boring nature.

Every time the notorious "Coke machine bandit" escaped with our money, we would come up with some sort of countermeasure to thwart his next attempt. None of the security precautions we employed did any good, though. At one point, we encased the entire machine inside a metal cage, but even that didn't stop him.

One morning, after returning from a late-night call at around 6:00 a.m., Stef went back to bed and left J.R. alone in the dayroom to finish the medical paperwork. There was only one more hour left in our shift, so I decided to violate one of my personal night-shift tenets and retire to one of the two bedrooms for a nap of my own. I figured the odds that we would be dispatched again before the shift ended were remote at this point. Even if we did get another call, it would be after sunrise, so I wasn't really concerned that I wouldn't be alert enough to fly if the pager went off.

About forty-five minutes into my nap, I was startled by the sound of someone banging on the door to the crew quarters. I figured someone from the oncoming crew must have forgotten his keys, and I knew J.R. was awake in the dayroom, so I didn't worry about it and tried to continue my nap. This proved to be impossible, however, because the banging from down the hallway continued, becoming louder and more annoying with each rap on the metal door.

Stef was irritated as well, and from the adjacent bedroom, I heard him yell at J.R.

"Open the damn door!"

The knocking stopped for just a few seconds, then started right up again, this time even louder.

Stef angrily rolled out of his rack and, without stopping to put on his boots, made his way down the hallway to see why J.R. was refusing to open the door. When he got to the dayroom, Stef was surprised to see that J.R. was nowhere in sight. He opened the door to see who was banging on it, and there was J.R., *clearly* agitated, with a man whom Stef had never seen. What was anything *but* clear to Stef, however, was why J.R. was holding the man in a headlock.

"It's the Coke guy!" J.R. shouted.

Still groggy from sleep, Stef was confused. He thought the term "Coke guy" meant that the stranger J.R. was rudely holding in a headlock was an employee of the Coca-Cola Company. This made absolutely no sense to Stef, who told J.R. to "let the poor man go so he can service our machine and be on his way."

"No! He's the bandit!" J.R. screamed indignantly. "Do you think it's okay if I hit him?"

"I don't think that's a good idea," Stef answered.

"Well, then get out here and help me hold him!" J.R. yelled.

"Hang on. I have to go get my boots."

With that, Stef closed the door and made his way, still half-asleep, back down the hallway to retrieve his boots from the bedroom. By the time he returned to help subdue the prisoner, the Coke machine bandit had escaped J.R.'s hold and disappeared down the street.

By the time our next transition night rolled around, and Stef was describing the incident to everyone at the Chili Parlor, the notorious Coke machine bandit was missing an arm, much like the mysterious and elusive one-armed man who had murdered Dr. Richard Kimble's wife in *The Fugitive*. Listening to him spin his yarn, I was quite proud of my young crew chief. He had mastered the art of embellishment, a skill that is critical to the telling of a good sea story.

Of course, I had no way to know whether he was telling the truth or not. I never got up to see what was going on that morning.

Off to the Sandbox

It goes without saying that December 7th is not just any other day to those of us who have served in the United States Navy. It was on that day in 1941 that planes from six Japanese carriers bombed, torpedoed, and strafed the U.S. Pacific Fleet at Pearl Harbor, sinking every ship that was moored along "Battleship Row," most notably, the USS *Arizona*. They also attacked Hickam Field and several other military installations on the island of Oahu, prompting America's entry into the Second World War. December 7th also happens to be Rose Caputo's birthday, and just two weeks following Rose's early morning rescue in 2004, the

Leper Colony was wrapping up "Pearl Harbor Day" at the Texas Chili Parlor.

If you'll remember, I mentioned this in the previous chapter. That was the same night that Scott "Zoob" Zublin, after learning we were the crew who had rescued his friend, congratulated us and comped our tab. What I failed to mention was that, prior to graciously picking up our tab that night, Zoob, while seated at our table, placed a phone call—to Rose Caputo. He wished Rose a happy birthday, told her how happy he was that she had survived her recent ordeal, and went on to tell her that her rescuers just happened to be regulars in his bar.

After learning that the STAR Flight crew that had appeared out of the fog to rescue her that morning was sitting across the table from her friend, Rose asked Zoob to pass the phone around to Stef, J.R., and me so she could tell each of us how grateful she was to be alive. That was the first time Stef and I had gotten a chance to talk to Rose Caputo. Up until that night, none of us even knew who she was. Of course, she and J.R. had already exchanged pleasantries during a brief, albeit loud, conversation on the end of the short-haul line—beneath our helicopter.

Fast-Forward to One Year Later

Because 2004 was a leap year, the Leper Colony's schedule, which normally would have advanced by one day from the previous year, fell on the same dates in December of 2005 as it had in December of 2004 (I could explain it to you, but it would only make your head hurt). At any rate, we were enjoying our second consecutive December 7th transition night, a phenomenon that wouldn't occur again until 2008 (trust me), and just as he had done the previous year, Zoob called Rose Caputo to wish her a happy birthday. He told her the Lepers were once again gathered around the table at the Parlor, and Rose, of course, wondered how it was that our transition night had fallen on December 7th two years in succession. . . . Sorry, remember how I told you in the second chapter that you were entitled to believe *most* of what I write in this book?

Moving on.

Just as she had done the previous year, Rose asked Zoob to pass the phone around so she could talk to each of us. This year, however, she was only able to express her gratitude to Stef Maier and yours truly. J.R. Esquivel wasn't at the Texas Chili Parlor that night. He was serving with the United States Army—at a forward operating base somewhere in the Afghan desert.

Following the terrorist attacks on September 11, 2001, J.R. had signed up for the Army Reserve, and because he held a nursing degree, he entered the Army as a first lieutenant, one grade above the normal entry level for newly commissioned officers. That, in and of itself, didn't make him any different from a lot of other young men who had wanted to serve their country after 9-11. The thing is, J.R. wasn't exactly a *young* man when he'd volunteered for military service. By the time he was finally called to active duty, in 2005, he was thirty-nine years old. The fact that he was deploying with men much younger than he was made his story all the more remarkable.

By then a captain, J.R. was deployed with a civil affairs unit. He and his fellow soldiers were tasked with building a road in southern Afghanistan, between Kandahar and Tarin Kowt. Strategically important, the road was subject to frequent IED attacks and ambushes by Taliban insurgents, placing J.R. and the other members of his unit in harm's way on a routine basis. On more than one occasion, J.R. distinguished himself while under fire, which came as no surprise to those of us who knew him.

J.R.'s next assignment was in the mountains near the Pakistan border, where he and his men worked closely with a Special Operations unit. It turns out that when the Spec Ops guys learned of J.R.'s experience as a flight nurse, they figured he'd be a valuable asset to them on their trips outside the wire. They reasoned correctly that, while flying for STAR Flight, J.R. had acquired a lot of experience taking care of trauma victims in the field. Realizing that these skills might prove highly valuable to them, they routinely shanghaied J.R. for many of their higher-risk missions, many of which were airborne assaults from a helicopter.

You'd have to know J.R. to understand that this was like throwing Brer Rabbit into the briar patch. He lived for this kind of stuff—and when he finally returned from his year-long deployment to Afghanistan, James Richard Esquivel was the recipient of a Bronze Star, awarded for acts of heroism and meritorious service in a combat zone.

Oh, and by the way, remember those Air Medals J.R. was convinced the Leper Colony was going to receive for rescuing that fallen rock climber from the ledge on Packsaddle Mountain?

You guessed it—he came back with one of those as well.

11

The Leper Colony Anthology

(Volume Three)

At the risk of sounding like I had a man crush on the guy, J.R. Esquivel's deployment had a significant impact on my psyche. It wasn't just for the obvious reason that I was losing a trusted crew member, one on whom I had come to depend over the previous five years. I struggled privately with the fact that he was leaving his family to fight in Afghanistan, and I was back here at home, living the good life. I left the Navy in 1992, not long after the First Gulf War had ended. Now here we were, a little over a decade later, back at it again. But as much as I admired J.R. for packing up and heading out to fight at thirty-nine years of age, I was pushing fifty, and the Navy wasn't looking for fifty-year-old combat pilots.

Still, I felt like I needed to do something to contribute to the war effort, so I spent the year J.R. was in Afghanistan maintaining his yard and helping to look after his family on my days off. While he was off fighting the Global War on Terrorism, I was in his yard every week, fighting the War on Weeds and Grub Worms. And even though firefights and IEDs were not part of my war, it was not fought entirely without risk.

There was a large oak tree in J.R.'s back yard, and it had a very solid, low-hanging limb attached to it. It was just the perfect height, too. As I pushed the lawnmower, looking down to maintain my track, the limb was hidden behind the bill of my baseball cap. As a result, I kept smacking it with my forehead. I must have done this a dozen times, and every time it happened, I would see stars and swear vociferously, unable to comprehend how I could keep running into the same limb—over and over again. I didn't know it at the time, but my doctor later told me that these micro-concussions, along with the dozens of others I had sustained in my lifetime, all contributed to the neurological deficits that ultimately led to the end of my flying career.

---

*Any club that would have me as a member,
I wouldn't care to join.*
—Groucho Marx (comedian)

Trying to figure out a way to stop pummeling myself senseless was not the only problem that had to be solved while J.R. was away in Afghanistan. His departure had left a vacancy in the Leper Colony. Initially, the empty slot on the flight schedule was handled on an ad hoc basis, sort of a "Leper-for-a-day" operation.

During the time our flight-nurse/rescuer position was a revolving door, Stef Maier and I were able to introduce many of our STAR Flight colleagues to our transition-night tradition at the Texas Chili Parlor. It was interesting to watch how different individuals, all from other crews, reacted to this ritual during their temporary assignments to the Leper Colony. Some of them were on board with the idea and even lamented that they didn't feel the same sense of comradery while flying with their regular crews. Others were not so enthusiastic. When the shift ended, they just wanted to go home and think about something else. It seemed that not everyone at STAR Flight relished the idea of joining our crew of misfits. I guess you can't blame them. Given our sometimes rocky relationship with the front office, hanging with the Leper Colony probably wasn't generally perceived to be a career-enhancing move.

One thing was certain, though. If the newest Leper was going to fit in with the rest of the Colony, the person who would wind up replacing J.R. Esquivel was going to have to enjoy transition night at the Chili Parlor—even if he had to fake it. Fortunately for us, someone came along who was perfect for the job.

The "Fourth Musketeer"

Just as the character d'Artagnan was to *The Three Musketeers* in the Alexandre Dumas novel, Howard Polden was not a replacement, but rather an addition to STAR Flight's Leper Colony. For those of you who are *Three Stooges* enthusiasts, I had originally planned to use a "Curly-to-Shemp" analogy, but I think the "Musketeers" comparison works better. I'm sure J.R. and Howard would agree.

A well-educated and highly skilled flight nurse, Howard bore a slight resemblance to Frank Sinatra—that is, if you can picture Sinatra as a slightly built Jewish guy from Yonkers. Although he didn't possess J.R.'s athletic stature, he was as energetic as they come. The kind of guy who was always willing to give you the shirt off his back, Howard was a team player and the perfect solution to the Leper Colony's manpower shortage. Just as I had with Stef and J.R., I quickly came to think of him as a brother rather than a co-worker.

Like J.R. before him, Howard came to STAR Flight from another flight program and hit the ground running. He was originally assigned to us just because that's where the empty flight-nurse slot happened to be, but it didn't take Stef and me long to realize we had stumbled into some good fortune when the front office permanently assigned him to the Leper Colony. Our bosses probably thought that Howard, whose character was above reproach, might just be able to keep Stef and me out of trouble. Imagine their disappointment the first time they saw him leaving the crew quarters wearing a wrinkled, half-buttoned Hawaiian shirt on his way to the Texas Chili Parlor.

In addition to fitting in as the Leper Colony's new flight nurse, it didn't take long for Howard Polden to prove his worth as STAR Flight's newest rescuer. His predecessor, J.R., was barely over the horizon when we rappelled Howard into the nighttime darkness of the rugged Barton Creek Greenbelt in April of 2005.

We were dispatched to search for two hikers who had gotten lost that afternoon. It was getting close to midnight (I told you these kinds of missions always come at night) by the time the first responders, searching on foot, finally decided to call us in. One of the hikers had a cell phone, and had been talking to the 911 dispatcher for several hours until the battery finally died. Based on that phone call, we knew one of the hikers had fallen from about fifteen feet and likely had a broken femur, which could potentially be a life-threatening injury.

When we finally located the two hikers, it was easy to see why the rescuers on the ground hadn't been able to locate them. They were in some of the thickest brush, in one of the most inaccessible areas in the entire greenbelt.

Because we knew one of the hikers had a broken leg, we wanted to get Howard to him as quickly as we could. Instead of taking the time to land and rig the short-haul, we elected to send him down on the rappel line. Howard quickly treated and packaged the injured hiker while Stef and I left to find an LZ so we could rig the aircraft for the rescue.

When we returned, Howard was ready to bring him out on the short-haul. We lifted the two of them and flew to the LZ, where we set them down and discussed what to do about the remaining hiker, who was still stranded in the greenbelt.

We were only a few short minutes from the hospital, so because he was injured, we decided to transport the first hiker to Brackenridge, then return for his partner. Stef radioed the rescuers on the ground and passed along the latitude and longitude coordinates where we'd left the remaining lost hiker. Based on that information, they estimated that, hiking through the rough terrain, they were still a good forty-five minutes from his position.

Because of the darkness and the rough terrain, we knew that, assuming they could find the remaining hiker, they would still have a difficult time extricating the exhausted man from the greenbelt. We told them to continue heading toward the stranded hiker (just in case we encountered a problem), but to anticipate that we would have him extracted before they could get to the scene.

Just as we had predicted, they were still twenty minutes away when Stef radioed the rescuers on the ground one last time and told them they could cancel their search. Howard had successfully rescued the second hiker, and we were calling it a night.

After short-hauling the second hiker from the greenbelt that night, Howard Polden had flawlessly executed his first two rescues as a member of the Leper Colony. Having wasted little time emerging from J.R. Esquivel's shadow, Howard was now a Leper in his own right—maybe not a big deal to most people, but a huge deal to Stef Maier and yours truly.

Just a few nights after we had pulled the two hikers from the Barton Creek Greenbelt, Howard was passionately regaling our fellow Chili Parlor patrons with the story of his rescues. There was no doubt about it. Howard Polden was the perfect man to fill J.R.'s shoes.

But if Howard thought he had a story to tell that night, he would learn just how simple those first rescues had been when, a couple of years later, he wound up on the business end of one of the most difficult short-hauls of my career. In the meantime, however, Howard Polden and Stef Maier were about to collaborate on a different kind of rescue—a *medical* rescue—and this one was nothing short of supernatural.

THE LEPER COLONY ANTHOLOGY (Volume Three)

"This actually 'was' in the brochure when I signed up."

If you fly an EMS helicopter, you'll occasionally witness a miracle. It happens about every twenty years or so. When I decided to make STAR Flight my career after the Navy, I hoped I would be able to continue doing what I loved—and in the process, maybe do something worthwhile as well. In the very first chapter of this book, I stated that I have been fortunate because, unlike most people, I never had to hold down a real job. I *lived* to fly. When I joined STAR Flight, I guess I was also hoping that, someday, someone else might live *because* I fly.

Well, a young eighteen-year-old man named Jacob Brochtrup lived. And while it may be true that all I did was drive the helicopter that day, I at least delivered the two guys who actually did save his life. I might not have been Jacob's guardian angel—but when literally every second counted, I was happy to provide the wings for those who were.

It was 12:19 p.m. when the pager went off on the first day of July in 2005. I just happened to be sitting in the aircraft, entering some new navigational waypoints into our GPS receiver, when the call came in as a boating accident on Lake Austin, next to Emma Long Park.

It was STAR Flight policy that we had to wear our survival vests on every flight, but mine was downstairs in the crew quarters. I briefly considered going down to get it, but my helmet was hanging beside me in the aircraft, and that was really all I needed. I was already sitting in the cockpit with the auxiliary power cart turned on, so I went ahead and started lighting the engines. By the time I got number two started, Stef and Howard had emerged from the stairwell, and Stef was carrying my vest. Howard unplugged the power cart and wheeled it away as Stef handed the vest to me and climbed into the copilot seat. Instead of stopping to put it on, I tossed it in the back of the aircraft and continued my run-up.

By 12:23, we were rolling final to the LZ at Emma Long Park. We had gotten airborne in less than a minute, and it had taken us a little more than three minutes to cover the seven miles to the park. That was good even by Leper Colony standards. In fact, I'm fairly certain that was the fastest response time we ever posted. It didn't look like it was going to be good *enough*, though. As we were touching down, I was witness to a grim sight. As several people, including two Austin EMS paramedics, were pulling a small boat ashore, I saw one of the medics administering CPR to a lifeless body.

"Well, that's not good," I said to Stef and Howard before I had even finished lowering the collective. "This guy's dead."

I guess it sounds inappropriately morbid now, but I used to joke that when we arrived on a scene, I could always determine who was faking and who was deceased. If the patient's true condition was anything between those two extremes, I deferred to the medical experts. In this case, Jacob's leg had gotten caught in a ski-boat propeller, and, barely attached by some skin at the hip, the severed limb was grotesquely hanging perpendicular to his torso. In addition to that, he was white as a ghost from having completely bled out through the gaping wound where his leg had been. You really couldn't blame me for misdiagnosing this one.

"Keep it hot!" Stef shouted on his way out of the aircraft.

Not surprisingly, Stef and Howard elected not to rely on the premature death pronouncement delivered by the only member of the crew who possessed no formal medical training. They spent the next few minutes working feverishly to save Jacob Brochtrup's life. After quickly loading him into the aircraft, they continued performing CPR on Jacob and kept him breathing on the short flight back to Brackenridge Hospital. They did this after we had landed as well, all the way into the ER, where the trauma staff was waiting for him.

I was still convinced it was a lost cause. Fortunately, I was just the driver that day, and my skepticism didn't matter to the people making the medical decisions. I had given up on Jacob—but Stef and Howard hadn't. Neither did the doctors and nurses who treated him in the emergency room and, later, in the intensive care unit.

Three days after we had transported him, on the Fourth of July, Jacob opened his eyes. Four days after that, doctors removed the ventilator that had been sustaining him since he'd arrived in the ICU. He was able to breathe on his own and even managed to say "hi" to his mom in a feeble, raspy voice. On July 13, Jacob was moved out of the ICU. An MRI showed that, miraculously, there was no damage to his brain—this despite the fact that he had lost almost every ounce of his blood on the day of the accident. On July 19, his doctors released Jacob from Brackenridge Hospital.

On August 1, 2005, I watched as Jacob walked, on a pair of crutches, into the room where my crew and I had assembled with all of the doctors, nurses, and first responders who had taken part in his rescue.

"You all saved my life," Jacob said with tears in his eyes. "I'm happy to be here."

I was happy he was there, too. . . . What's more, I was happy *yours truly* was there to see it.

And Igor Sikorski would have been happy to see how my crew and I had put his invention to good use that day when Jacob Brochtrup, with help from Stef Maier and Howard Polden, cheated death in the back of that STAR Flight helicopter. It had been roughly five minutes from the time the call had come in until Stef and Howard laid hands on Jacob as he was being pulled on shore at Emma Long Park.

There were a lot of unrelated, individual circumstances that had necessarily aligned in perfect symmetry in order to make the unprecedented—not to mention, never-again repeated—response time possible. The medical professionals who treated Jacob had never seen this type of recovery from such a traumatic injury. By all rights, he should have died on scene. So, was it a miracle of divine intervention? I can only speak for myself, of course, but if you choose to believe it was, I seriously doubt you'd get an argument from anyone who witnessed it.

A few weeks later, Stef, Howard, and I got permission from the front office to take Jacob for a ride in the same helicopter we'd used to transport him to the hospital. He seemed to be enjoying the flight, and he asked if we could do any "tricks." In a move that would have surely drawn the ire of my bosses, I chopped the throttles to idle and demonstrated an autorotation for him. I figured, after what he'd been through, he deserved a good demonstration of our helicopter's "engine-out capabilities."

After I had rolled the throttles back up and recovered from the autorotation, he asked if I could show him "some more exciting stuff like that."

Not wanting to push it, I told him, "That's about as exciting as it gets when you're flying with the Leper Colony."

If he could have been with us a little less than two years later, he would have known I was not being truthful.

Bobbing for Kayakers

On June 28, 2007, a series of repeated, heavy rains transformed the Lampasas River, thirty miles north of Austin, from a gently flowing waterway into a raging torrent, the likes of which the residents who lived near the river had never seen. Two of those residents decided it would be a good day to test their kayaking skills. The result was a frenzied ride along miles and miles of debris-filled water as they clung to their overturned kayaks through parts of three Texas counties.

During the fifteen minutes it took us to fly to the scene, about a dozen first responders, dangling ropes from several bridges, tried desperately to rescue the two men, only to watch helplessly as they went barreling underneath them and emerged on the other side of the bridge. After each failed attempt, they hurriedly drove downstream to the next bridge and tried again. Each time they moved, they gave us updated GPS coordinates to their current position.

This went on for the duration of our flight from Austin, and when we finally arrived on scene, we were not encouraged by what we saw in the river. The uprooted trees that were traveling in the main channel, along with the breakneck speed of the current, were going to make this anything but a routine swift-water rescue (if there is such a thing).

Trying to get ahead of them, we quickly located an LZ a couple of miles downstream from the two kayakers and set up for an approach. In my haste to get down so we could deploy Howard on the short-haul before they arrived, I almost over-torqued our transmission. Coming straight down, I allowed my descent rate to get too high, and when I pulled hard on the collective to arrest it, the torque gauge shot all the way to the redline. Fortunately, we landed without damaging the airframe, but I decided, right then and there, that I needed to slow down. I had to make sure I didn't do anything stupid while trying to rescue these guys, and not just because they'd had no business getting themselves into this mess in the first place. My first responsibility was to my crew, and after thirty years in the cockpit, I shouldn't have needed to remind myself of that fact.

Stef and Howard jumped out to rig the short-haul, and while they were busy doing that, I raised the on-scene commander on the radio for an update on the position of the two kayakers. Based on what he told me, I knew they would be coming past us soon, but in light of the blunder I had just narrowly averted, I didn't relay this to Stef and Howard because I didn't want them to feel rushed.

After making sure Howard was properly secured to the short-haul line, Stef climbed back in and reconnected his comm cord.

"You got any ideas as to how we're actually gonna do this?" I asked him.

"Let's try putting Howard out in front of them and see if he can grab them as they come to him," Stef replied.

"Roger that. Does Howard know that's the plan?"

"Yep."

We got a thumbs-up from Howard and lifted him just high enough to clear the partially submerged trees that lined the flooded banks of the river. Once we got airborne, I was relieved to see the two kayakers hadn't already passed our LZ while we were rigging. They were coming toward us, about a thousand yards upriver. We had timed it just right, and as we gently lowered Howard into the raging deluge, the force of the current immediately carried Howard to the downriver side of the aircraft. This put a strain on the short-haul line, and I could feel it pulling against the helicopter as I tried to maintain our hover. With the starboard side (my side) of the aircraft toward the kayakers so Stef and I could both see them coming, I steadied our position.

At this point, we both had eyes on the targets as they rushed toward us, so had I chosen to, I could have relied on my own intuition to position the helicopter in their projected path. Instead, I trusted Stef to line us up. I had depended on him for almost eight years to this point, and I didn't see any reason not to trust him this time. Besides, I still couldn't see the short-haul line, and Stef could. One of the kayakers was about a hundred yards or so in front of the other, and Stef zeroed in on him.

"Forward three," he said.

Because we were hovering ninety degrees to the channel, he was actually moving me left and right with his "forward" and "back" commands. A less-experienced crew chief might have struggled with this at first, but Stef handled it like it was something we did all the time.

"Back one," came the next command. "Dammit!" Stef shouted a few seconds later.

We had missed the first kayaker.

Stef quickly tried to reposition me for the trailing kayaker, but there wasn't enough time.

"Let's get out in front of them again," he said.

As quickly as we could do it without slinging Howard around like a jumper on the end of a bungee cord, we lifted him from the water and carried him over the trees, past several bends in the river. Once again, we stationed ourselves in front of the kayakers and waited.

"Come down a little," Stef told me.

He was trying desperately to give Howard the slack he needed to maneuver in the water, but it was hopeless in the rapidly moving current. As soon as we descended, Howard was immediately swept to the end of the short-haul line, and it became taut once again. There was no way Howard could move back and forth without slack in the line, and we missed both kayakers for a second time.

We raced to get in front of them again, and—undeterred—Stef lined me up as we waited to give Howard one more chance to snag our moving targets. Once again, Howard was immediately swept to the end of the short-haul, which left him no chance to maneuver. Just as he had the two previous times, the lead kayaker went rushing past Howard, just out of his reach. This time, however, Stef was able to put Howard in a perfect position to intercept the trailing kayaker, who was closing fast.

"Dammit!" Stef shouted again.

He didn't have to tell me what had happened. Thanks to the tension on the short-haul line, I could feel the kayaker slam into Howard with the force of a sledgehammer. Just as quickly as I had felt the additional weight of the kayaker on the short-haul, it was gone again. Even if the two targets had not been desperately clinging to their kayaks, which were essentially battering rams at that speed, there was no way Howard could stop that much momentum while working from a stationary position. Even if we could signal them to let go of the kayaks, trying to stop them was still going to be too dangerous—for Howard as well as for the kayakers.

"This is no good," Stef said dejectedly over the ICS.

At this point, I became concerned about the beating Howard was taking in the water, but Stef said he had signaled up that he was okay.

"What if we try pacing them?" I asked Stef.

"That may be the only way Howard's gonna be able to hold onto 'em," he answered "I already know doin' it this way's not gonna work."

"Do we need to land and brief Howard?" I asked him.

"Howard's a smart guy. He'll figure it out."

Again, we lifted Howard out of the water and raced the kayakers downriver. This time, however, instead of stationing ourselves well out in front of them and waiting, we lowered Howard just a few yards ahead of the kayakers while trying to match their speed.

This proved to be more difficult than we had anticipated. On the first few attempts, I tried sliding sideways down the river, starboard side to the kayakers, so I could keep them in sight like I'd done on our initial attempt to snag them. It didn't take me long to decide this was a bad idea. It was too hard for me to watch for obstacles downriver while looking at the kayakers, so I kicked the nose straight downstream and, once again, trusted Stef to tell me what I needed to do.

Several times, while pacing the kayakers, we were forced to break it off and lift Howard over bridges and power lines. Stef was getting frustrated because this always seemed to be necessary just as Howard was about to make contact with one of them. On two different attempts, just as he was ready to give me one last command to get Howard into position for the rescue, he heard me call "Bridge!" and up we went, forcing him to start the whole tedious process again.

During this entire affair, the first responders were racing along the road next to the river, trying to keep up with us. If it hadn't been a life and death situation, I'm sure it would have been comical to watch. A couple of times, they had stationed themselves on the next bridge in front of us, and Howard probably could have shaken hands with them as he flew past like Peter Pan, suspended on the short-haul line.

Finally, we came around a bend in the river and saw a welcome sight. It was a long, unobstructed straightaway, which allowed us a brief respite from the obstacle course of bridges and power lines we'd been dodging while trying to fly formation on the kayakers. Seizing the opportunity, Stef quickly had me lower Howard just in front of the lead kayaker as we were pacing him. As soon as Howard's feet were in the water, precisely where Stef wanted him, my seasoned crew chief wasted no time.

"Down ten feet now!" he shouted.

This finally gave Howard the slack he needed to maneuver his way through the swiftly moving water and get the rescue ring around the first of the two kayakers. The kayaker let go of his capsized boat, and Howard gave Stef the signal to lift.

Because we were all moving at the same speed as the current, there was no relative motion between the helicopter and the people in the water. This meant there was still a lot of slack in the short-haul line. Stef carefully took in the slack to keep the line from fouling in the abandoned kayak, along with the rest of the floating debris that was accompanying Howard and the man on their race down the raging river.

"Easy up," Stef said calmly. "Tension's coming on now."

I felt the aircraft take the load, and without waiting for what I knew was the next command, I raised the collective and started a slow climb, still pacing the forward speed of the river. As soon as Stef told me our short-haul passengers were out of the water and clear of the trees, I peeled away from the channel and looked for a place to set Howard and the kayaker on the ground.

This would normally have been a predetermined DZ (drop zone), but because the rescue had begun well over ten miles upriver, we had to improvise. I told Stef the nearby road that ran beside the river looked to be our best option, so he immediately radioed the first responders, who had been following us along the road in emergency vehicles, and asked them to block it off for us.

As soon as the first kayaker was on the ground, Howard disconnected him from the short-haul and handed him off to the first responders. Looking down at the spectacle in the DZ beneath him, Stef keyed the ICS and chuckled slightly.

"This guy can't stand up, but I think it's just because he's exhausted," Stef said.

Stef watched the first responders escort the staggering, just-rescued kayaker from below the helicopter. As soon as they were clear, he saw Howard give him a thumbs-up.

"Oh well, I'm not sure if he's injured or not, but he's someone else's problem now. Easy up."

By the time we reached the second kayaker, he had traveled well beyond the clear straightaway we'd successfully used to our advantage on the first rescue. Once again, we resumed hurdling bridges and power lines while negotiating turns in the river, and once again, we were forced to abort several attempts because of obstructions in front of us.

Eventually, Stef was able to deliver Howard to the remaining kayaker, who was reluctant to let go of his ill-fated boat. By this time, Howard was in no mood to argue with the guy. Even though they were beneath the helicopter, where I couldn't see what was happening, Stef let me know that Howard had done a good job "convincing" his stubborn guest on the short-haul to let go of the kayak.

We repeated the road-blocking drill with the help of the first responders, who set up a second DZ, almost two miles downriver from the first one. More than an hour and a half after his ordeal had begun, the second kayaker was safely deposited onto a stretch of dry asphalt.

Just as he had done with the guy's buddy, Howard remanded him to the paramedics on the ground. This time, however, Howard's task was finished, and he disconnected himself from the short-haul line and sat down for a well-deserved rest. Stef pulled the now-empty short-haul line in, and we looked for a place to land, a little farther down the road.

A few minutes later, after we had landed and shut down, Howard came sauntering into the LZ to rejoin Stef and me. He had a wry grin on his face that was born of both fatigue and satisfaction.

"Just another day's work," he said in that unmistakable Yonkers accent.

Turning the Page

Howard Polden got it wrong that day in Lampasas. It hadn't been "just another day's work." That rescue turned out to be the last, and most difficult, of the twelve Stef and I logged with Howard. Counting the successful rescue missions we'd flown with J.R. Esquivel, Stef and I had logged more than thirty together. Several of those missions, two with Howard and three with J.R., had involved multiple rescues. All things considered, it had been a pretty good eight-year run. It seemed I had chosen wisely when, at the beginning of that run, I asked Lourdes Maier's kid to be my crew chief.

Just as it always does, however, time moved ahead, and the changes it left in its wake began to disrupt my comfortable STAR Flight existence. The third chapter in the Old Testament book of Ecclesiastes begins with these familiar words:

"There is a time for everything, and a season for every activity under the heavens."

It goes on to say, "So I saw that there is nothing better for a person than to enjoy their work."

I did enjoy my work, in a way that few people are ever fortunate enough to experience—but I was about to enjoy it just a little less. At the risk of sounding melodramatic, the Leper Colony's season was nearing an end.

STAR Flight was expanding. In 2006, we had traded our two EC-135 helicopters for two larger EC-145s. A little more than a year later, shortly after the Lampasas River rescues, we equipped them with rescue hoists. Not long after that, a third EC-145 was added. During this process, additional flight crews were hired to staff a second on-duty aircraft. For every pilot that was hired, a total of two medical crew members were hired to fly with them—and Stef Maier was tasked with a large portion of their training. We essentially became the training crew for the new medics and nurses coming to STAR Flight.

Additionally, the rookie pilots who were being hired needed to be battle-rostered with experienced medical crews. Following that final rescue mission in June of 2007, Howard Polden continued to fly with Stef and me for another six months before being reassigned to fly with one of the newly hired pilots. Stef continued to fly with me for several months after that until, finally, he was reassigned to a new pilot as well.

The Leper Colony was permanently disbanded.

Societal mores and outward appearances aside, give me the opportunity to watch a man take on a difficult task—day in and day out, while under inconceivable pressure—and I'll accurately assess the strength of his character in short order.

—Kevin McDonald (STAR Flight pilot)

I can't begin to explain how much those guys meant to me. For eight of my twenty years as a STAR Flight pilot, I had the perfect job, flying with the perfect crew.

J.R. Esquivel was a true hero, and it wasn't just because he had earned the Bronze Star and an Air Medal while serving in Afghanistan. He might have hidden it well from most people, but

I had always known that J.R. was one of the most compassionate and committed human beings on the planet. During the time we flew together, I watched him mature (from being a bit of a carouser) into an exemplary family man, a model citizen, and a true leader—the kind people look to when things get tough. That said, even though he's now a well-respected and properly domesticated father of two, I'd still pick J.R. to be on my side in a bar fight.

After J.R. left for Afghanistan, Howard Polden was a godsend. He was Jewish, and I was a Christian, but we were kindred spirits, and I came to love him like a brother. An avid equestrian, Howard and his wife now own a beautiful spread of property north of Austin, and we're still good friends to this day. I like to tell Howard that, of all the Texas ranchers I know who grew up on the streets of Yonkers, he's my favorite.

And then there's Stef Maier. All totaled, he and I flew more than thirteen hundred missions together, and on every one of them, he validated my faith in his crew-chiefing talents. For a large portion of my career at STAR Flight, he helped to keep me safe. He was one of the reasons it was fun to go to work every day. Far from being the ne'er-do-well I had pegged him to be on that first day we met, Stef turned out to be the glue that held the Leper Colony together for all of those years.

And, as for the "Leper Colony" moniker itself—well, in the movie *Twelve O'clock High*, that maligned crew of flying misfits ultimately redeemed themselves in heroic fashion. As the curtain comes down on the STAR Flight version of that story, I look back on the significance of those eight years together, and, considering the positive impact we made, I can't help thinking—my crew of flying misfits managed to do okay as well.

Our meeting with Jacob Brochtrup, exactly one month after his near-death experience on Lake Austin—As I greet Jacob, Howard Polden (middle) and Stef Maier (far right) look on. They, along with the two Austin EMS paramedics pictured here, Anthony Flood (far left) and Craig Fairbrother (2nd from right), as well as the Brackenridge Hospital ER doctors and nurses, were the real heroes that day. *Photo courtesy of Austin American Statesman*

Jacob Brochtrup with the second-generation Leper Colony on the day we took him flying with us in the EC-135—Left to right (in flight suits) are yours truly, Stef Maier, and Howard Polden.

Hoist training in the EC-145.

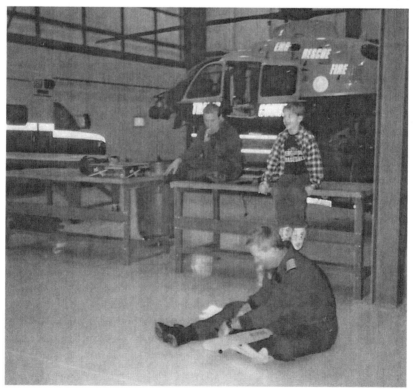

Christmas morning at the STAR Flight hangar—Stef Maier looks on as my son, James (seated next to him on the work table), anxiously awaits the maiden flight of his brand-new gasoline-powered Piper Cub control-line model. The flight ended tragically when, after several circuits around the empty hangar bay, I clipped a wing on the I-beam, which can be seen just to the right of the table. Fortunately, that was the only airframe I ever bent.

12

Ejection Seats, Fires, and Flameouts

Mistakes are inevitable in aviation, ... The trick is not to make the mistake that will kill you.

—Stephen Coonts (author)

On July 17, 1978, as I prepared to make my first solo cross-country flight from the Arlington (Texas) Municipal Airport, I inadvertently walked into the trailing edge of the wing on my rented Cessna-150 and put a three-quarter-inch gash in my forehead. That wasn't the first concussion in my life, and—as you already know—it wouldn't be the last. As humiliating as it was to crack my head open on the wing of that Cessna, it definitely wasn't the dumbest mistake I would make over the course of my flying career.

In my defense, I did manage to log more than eleven thousand hours without so much as bending an airframe, but I also made my share of gaffes along the way. Some were errors in judgment, while others were the result of complacency. Some were hybrid mistakes—poor judgment, caused by complacency. The important thing is that I managed to survive them all, and now that I'm retired and no longer subject to disciplinary action from the United States Navy, Travis County, or the Federal Aviation Administration, I'm finally free to confess my aeronautical sins without fear of reprisal.

Because I'm not Catholic, and we Protestants don't do confessionals, I can only guess that this is what it feels like to spill your guts to a priest on the other side of a curtain. At any rate, I'm about to get some things off my chest here, and perhaps there's a young aviator out there somewhere who might benefit from it. For the rest of you, I've included these stories purely for their entertainment value.

The Martin-Baker Fan Club

If there's one thing the Navy was big on, it was checklists. In theory, the checklist was there to prevent aviators from making mistakes, but as Forrest Gump famously pointed out, "Stupid is as stupid does." Roughly translated, that means no checklist is pilot-proof.

I learned this one day early in my career, when I was still a student naval aviator. I was fat, dumb, and happy, going through the run-up checks in my T-28. For those of you who aren't aviation enthusiasts, pilots do run-ups prior to takeoff, just to

EJECTION SEATS, FIRES, AND FLAMEOUTS

ensure everything is working properly at full power. The T-28, you may remember, was equipped with a 1425-horsepower radial engine, which sounded quite impressive at full power; and I loved hearing that thing roar whenever I pushed the throttle to the firewall. However, the thing about radial engines is, they have different power limits for different RPM settings. If these power limits are exceeded, you can damage the engine because of a phenomenon known as "overboosting." This is actually true for all reciprocating aircraft engines that power a constant-speed propeller.

I should know. That's exactly what I did one day when I skipped over the step in the checklist where you're supposed to advance the propeller to the maximum RPM setting before advancing the throttle to full power. And therein lies the problem with checklists. Unless you have a copilot to read the thing to you in a "challenge and reply" fashion, you have to take your attention *away* from the page in order to complete the tasks prescribed *on* the page. You're more likely to skip a step when you look back down at the checklist than if you had just memorized the procedure in the first place.

At the risk of sounding like a heretic, this is exactly why I've always maintained that checklists are useless without a copilot. Do it the exact same way every time, over and over, until you can do it in your sleep. You'll be far less likely to screw it up than if you depend on a checklist that you have to read by yourself.

Sometimes you don't even need to skip part of the checklist to screw it up. I was taxiing back to the ramp after landing my T-2 Buckeye one day and very nearly took a ride in the ejection seat. As part of the after-landing checklist, you were supposed to replace the safety pins in two large handles (one above your head, and one in front of the seat pan), either of which—if pulled—would fire the ejection seat and send the pilot merrily on his way out of the aircraft. I got the top pin inserted just fine, but because I couldn't get the lock on my shoulder straps to release, I had a hard time bending forward to replace the one between my knees.

Without thinking (that should always be a tip-off that you're about to screw something up), I tried to grab the front edge of the seat pan to pull myself forward. In the process, I inadvertently grabbed the still-unpinned handle (the same handle I was trying to pin so I couldn't accidentally eject myself) instead of the seat pan. The motor-skills part of my brain told me to pull so that I could lean forward to insert the pin, but the reasoning part of my brain vetoed that action—just in time to keep me from joining the Martin-Baker Fan Club, the name given to the list of aviators who have taken a ride on the Martin-Baker Aircraft Company's ubiquitous, rocket-powered seat.

It was a "zero/zero" seat, meaning you could eject while your aircraft was on the ground, stationary, and the parachute would deploy just in time to give you one good swing before you hit the deck. I'm pretty sure I would have survived, but I'm also sure that had I ejected myself from a perfectly good T-2 while taxiing it back to the flight line, it would have prematurely ended my career as a naval aviator.

Now, it's one thing to inadvertently skip a step while you're using a checklist (or even to make a mistake while following the checklist)—it's another thing altogether to just skip the checklist entirely. That's what my old buddy, Harold Graebe, and I did one day as we were getting ready to land our SH-60 *Seahawk* on a remote island in the middle of the Indian Ocean.

EJECTION SEATS, FIRES, AND FLAMEOUTS

"Who set the parking brake?"

While our battle group was preparing to head up to the Persian Gulf for Operation Earnest Will in 1987, the cruiser on which we were deployed (USS *Valley Forge*) was taking on supplies at the U.S. Navy base in Diego Garcia. A few days earlier, while it was still at sea, Harold and I, along with two other pilots from the *Valley Forge*, had flown our two helicopters from the ship to the airfield on the island.

The ship's captain had never been up in a *Seahawk*, and he asked me to take him for a "dog-and-pony-show" flight. We drove to the airfield and mounted up for what we thought was going to be a leisurely sightseeing sortie.

Halfway through the flight, however, we were contacted by the ship's operations officer, who told us the captain needed to report to the USS *Constellation* (the aircraft carrier assigned to our battle group) for a high-level confab with the battle group commander. Normally, this would have involved taking a small boat from our pierside cruiser to the carrier, which was anchored in the middle of the harbor. Because we were already airborne, however, the skipper asked if I could fly him directly to the *Constellation* instead.

I was already turning back toward Diego Garcia, and I told him, "Sure thing, Skipper, but I'll have to land at the airfield and pick up Lieutenant Graebe before we fly out to *Connie*."

The SH-60 NATOPS manual required a body in the copilot seat, so in order to fly back to the airfield, I would need Harold to take the captain's place once we landed aboard the

249

carrier. When I explained all of this to the *Valley Forge* operations officer, he asked me if I knew where Lieutenant Graebe could be found, and I told him he would most likely be asleep in his rack or watching movies in the ship's wardroom.

Naturally, I was correct.

Harold was waiting for us when we landed at Diego Garcia, and he quickly jumped into the back of the aircraft. The plan was to have him move to the copilot seat after we landed aboard the *Constellation*. It all seemed simple enough.

Well, after delivering our captain, we were sitting on the flight deck of the carrier, and Harold climbed into the copilot's seat as planned. I gave him control of the aircraft for the short flight back to the airfield, and this is where the plan started to break down.

Harold was at the controls, getting his takeoff clearance from the air boss aboard the *Constellation*. Because we needed a landing clearance back at the airfield, and because the flight to the runway was only going to take about thirty seconds, I decided to be proactive. I got on the other radio to the tower at Diego Garcia and let them know we were about to lift from the carrier.

So there we were. Harold was talking to the air boss on one radio, I was talking to the tower on another radio, and neither of us was talking to the other guy in the cockpit. This turned out to be problematic because, in accordance with the landing checklist, I had set the parking brake when we'd landed aboard the carrier. Now, Harold, who had not been in the cockpit during the time I had completed the checklist and set the brake, was at the controls for the short hop back over to the airfield.

When we lifted from the flight deck of the carrier, we were almost immediately on final for a *running* landing to the runway on the island. I was the one communicating with the tower, so I told

Harold (who was still talking to the air boss aboard the *Constellation*) that we were cleared to land. Because we were both busy on separate radios, neither of us took the necessary time to go through the landing checks, which would have cued us to release the parking brake before we touched down *at 60 freakin' knots* (dammit!) on the concrete runway. We both recognized instantly what was happening, and we both pulled back on the cyclic and up on the collective at the same time, but it was too late.

The starboard main-mount never touched down, but the tire on the port side had exploded on impact. After hover-taxiing to the ramp, we had to stay airborne until we could raise our maintenance guys on the radio and ask them to come out to inspect the landing gear. While we were hovering over our parking spot, they concluded that, even though the tire was almost completely gone, neither the wheel assembly nor the strut had been damaged in the mishap. We waited there, suspended in shame, while they conspicuously brought out a mattress, rolled it up like a cigar, and held it in place so we could set our exposed wheel (the one that no longer had a tire on it) down without further damaging it on the concrete. Then we shut the aircraft down, and, with our heads down and our tails between our legs, we began the long march to the hangar to explain ourselves to the detachment maintenance chief. Luckily for us, our maintenance chief possessed a good sense of humor, and he let us off the hook without subjecting us to too much humiliation, even though we deserved it.

Even now, whenever we get together for a beer call, Harold still likes to point out that I was the aircraft commander on that flight, and therefore, I was responsible. I always counter to Harold that, while it *is* true that I was the aircraft commander that day, he was the pilot at the controls—just as he was the day we nearly ate a set of high-tension power lines in the Mojave Desert.

A Matter of Perspective

If you've ever wondered why helicopter pilots hate power lines, maybe this will help you to understand it a little better. During the time Harold Graebe and I were training for our deployment aboard the USS *Valley Forge*, we were on a high-speed, low-level flight east of Death Valley one afternoon. I was navigating, and Harold was flying, as we were screaming along at 160 knots, 50 feet above the desert floor.

I looked up from the chart and spotted a set of high-tension power lines on the next ridge, about a mile in front of us. I called "power lines," and Harold responded that he had them in sight, at which point I looked back down at the chart. When I looked up again a few seconds later, we were staring directly at the power lines and closing fast.

Just as we would later do in the Diego Garcia incident, Harold and I both pulled back on the cyclic at the same time. Our crew chief said afterward that he could have reached down from the cargo door and "played a tune on the wires" as we passed above them by the narrowest of margins. Harold had seen the power lines okay—but to him they appeared to be on a second ridge, just past the one where we'd nearly eaten the power-line sandwich.

Now, at this point, you're probably thinking to yourself that Harold and I were just a couple of screw-ups, but I'm telling you about these incidents to give you an idea just how quickly things can go south while you're flying a helicopter. Harold Graebe was, without question, the best pilot in our squadron. If he and I could nearly crash during a training flight in broad daylight, you should begin to appreciate just how dangerous the act of piloting a helicopter can be when you let your guard down—even for just a few seconds.

Aviation, in general, is inherently unforgiving. The list of competent, well-respected aviators who have died while plying their trade is sobering to contemplate. Unfortunately, it seems that, all too often, those of us who weren't there and, therefore, can't possibly know how we would have reacted to whatever emergency it was that ultimately did them in, are quick to condemn the pilots who don't survive and aren't around to defend their reputations. We like to think we would have managed the situation better than they did. We tend to forget that they didn't have the benefit of our 20/20 hindsight.

Fortunately, Harold and I both managed to survive our Navy flying careers, and after leaving the Navy, we both made the transition to commercial aviation. Of course, just because I had become a civilian pilot, it didn't necessarily lessen the day-to-day risks associated with flying. In fact, except for the fact that I no longer had to worry about night shipboard landings, flying for STAR Flight was often more hazardous than flying in the Navy had been.

When we were flying helicopters for the Navy, Harold Graebe and I had avoided flying inside the *dead man's curve* as much as we could. At STAR Flight, that's where I made my living. Almost every landing, it seemed, was a precision, steep approach. In addition, when we were fighting fires or conducting rescue operations, we were almost *always* on the wrong side of that curve—frequently in bad weather.

I still say it was better than getting a real job, though. I'll repeat what I said in the very first chapter: I can't imagine how different my life would have been had I been forced to earn an honest living. Flying for STAR Flight was worth the risk, if for no other reason, because of the people with whom I shared the risk. Just like the bond I had forged with Harold and my other squadron mates while flying in the Navy, the relationships I developed with my colleagues at STAR Flight were priceless to me. And it wasn't as if we were cheating death every time we showed up for work.

Even when things did go wrong, they sometimes turned out to be more humorous than perilous. That was certainly the case one night when I unintentionally disrupted the 3:00 a.m. smoking break outside the door to the Brackenridge Hospital Emergency Room.

"We've got a fire back here!"

Carl Shropshire was a bit of a paradox. A Vietnam veteran, he was part RECON Marine, part South-Austin hippie. He was also one of the most intriguing personalities ever to wear the STAR Flight uniform. Beneath the exterior of wire-rimmed glasses, the ponytail, and the colorful bandanas he always wore around the neck of his flight suit, Carl was a skilled paramedic. He also possessed the best bedside manner of anyone I ever saw during my twenty years at STAR Flight.

Carl, along with Mary Neely (a STAR Flight nurse during the early nineties), was working with me one night when we were still flying the Bell 412. We got a call for an inter-facility (hospital to hospital) transfer just before three o'clock in the morning. It was a hot, muggy night, and when we got upstairs, the usual crowd of nurses and medical technicians had gathered outside the emergency room, where they routinely lit up cigarettes and shot the breeze during their breaks (this was back before you got shipped off to Siberia for smoking in public).

Their venue, next to the sliding glass door that led into the ER, was about a hundred feet from the helipad, so watching us launch provided good entertainment for them while they satisfied their nicotine addictions.

Mary was sitting next to me, in the copilot seat, and Carl was next to the aircraft, manning the auxiliary power cart. As was my custom during night shifts, I started the number two engine first.

The first engine you engage tends to start a little hotter because the rotors aren't yet turning when it lights off and also because you're spinning the turbine with battery power instead of an onboard generator. By alternating which engine you start first, you put an equal amount of stress on both engines over time. Some pilots would alternate each time they started the engines, while others would alternate every other day. I normally started the number one engine on day shifts, and then I would start the number two engine on night shifts.

At any rate, after the number two engine started normally, Carl secured the power cart and strapped himself into an aft-facing seat in the back of the aircraft. When I tried to light the number one engine, which should have been the easier engine to start at this point, it wouldn't light off. Recognizing the failed start, I secured the starter and waited for it to cool down. Then I vented the engine (motored it with the fuel valve off) to clear any excess fuel from the combustion chamber and attempted another start.

Well, I must not have vented the combustion chamber quite long enough because, when I hit the starter again, we lit up most of downtown Austin. We literally blew a ball of burning fuel onto our tail boom, and the "ENGINE FIRE" warning light illuminated in the cockpit.

"We've got a fire back here!" Carl yelled over the ICS.

I pulled both emergency T-handles, which shut off fuel to both of the engines. This secured the number two engine (the one that had been running) and armed the fire bottles. As I was reaching overhead to discharge the first of two bottles into the number one engine bay, I caught a glimpse of Carl's backside in my peripheral vision. He was running toward the large, wheel-mounted fire extinguisher that was parked on the corner of the helipad.

That a boy, Carl! I thought to myself. *He's going for the fire extinguisher.*

After hitting the toggle switch that discharged the first fire bottle, I looked outside the cockpit again, and Carl was still running . . . past the fire extinguisher . . . and away from the helipad . . . just as quickly as he could! Just beyond Carl, I could see the 3:00 a.m. smokers fighting each other to get through the door leading back into the ER.

In the meantime, the fire had been extinguished, and Mary and I were the only ones left in the aircraft.

Mary looked over at me from the copilot seat and asked, "Is it okay for me to get out now?"

"Mary," I said, "the next time we catch on fire, you have my permission to un-ass the aircraft without delay."

She just stared back at me and grinned.

Fortunately, everything had turned out alright, and in retrospect, I couldn't really blame Carl for running away from the helipad. I honestly don't think it was the first time in his life he'd scrambled to escape from a hot LZ.

EJECTION SEATS, FIRES, AND FLAMEOUTS

I later told him he deserved a below-average in attitude for not coming back with the fire bottle, but I also told him he deserved an above-average in headwork for allowing his survival instincts to dictate his actions. When you don't have much time to think, the will to survive can often lead you to make the correct decisions in a life-threatening situation.

Surviving Inside the Dead Man's Curve

The original helipad at the North Austin Medical Center, NAMC for short, was not in a location I would have chosen back in August of 2000. The prevailing winds in Austin are from the southeast during the hot summer months, and there were a series of obstacles just south of the pad that made climbing out in that direction difficult. Takeoffs there were always a little tricky, and what was potentially worse, a pilot trying to wave off an approach during an emergency would be forced to negotiate a chain-link fence, a portable building (located right next to the pad), and a set of high-tension power lines just across the street.

We didn't fly to NAMC all that often, and they eventually moved the pad to a more desirable site, but that was a few years after I had already demonstrated, in near tragic fashion, exactly why the original location had been less than ideal.

Stef Maier was in the copilot seat that day. It was hot and humid (both of which degrade the aircraft's performance), and to make matters worse, we were unusually heavy, both with fuel and people. We were carrying an extra crew member for training purposes, and because the flight to the scene had been a short one, we hadn't burned much fuel en route. STAR Flight had recently hired a new nurse, and Stef was up front with me because the new nurse and his instructor (another medic) were in the back of the bird, treating the patient.

We were on short final to the NAMC pad at around 75 feet and 20 knots, well inside that ever-present *dead man's curve*, when, just as I was adding power to slow our closure rate to the pad, I caught a glimpse of several caution lights illuminating in rapid succession. Before I could look inside the cockpit to read them—we had lost the number one engine.

According to the performance specifications published by the good folks at Eurocopter, we should have crashed. Fortunately for us, their helicopter was better than advertised, and I lowered the nose—trading what little altitude I had for some airspeed—and managed to get our EC-135 flying again. As I accelerated to 40 knots and desperately struggled to arrest our descent and start a climb, we narrowly cleared the fence and the portable building.

Normally, turning the aircraft (which requires additional lift) would be the worst thing a pilot could do while he's trying to climb on one engine, but at the time, it seemed like a better option than flying into the high-tension power lines directly in front of us.

To this day, Stef claims I was praying out loud as we made that ninety-degree turn, trying desperately to escape an almost-certain fatal encounter with the wires. We managed to evade the power lines, but it was costing us dearly in altitude, and we didn't have much to spare. I was still way too low—and way too slow to climb. I don't remember talking to God (I probably should have been), but I do remember talking to that helicopter. With every beat of my palpitating heart, I was pleading with our EC-135, begging it to fly. For several seconds, we continued to bleed precious rotor RPM as we sank helplessly in the hot Texas air. Then—against all odds—the crippled machine began flying again!

I couldn't believe our good fortune and tried in vain not to laugh out loud. Once we had leveled off, I gradually accelerated to 70 knots, established a barely positive rate of climb, and regained my composure.

We had just dodged a major bullet, but we still weren't completely out of the woods. Because we couldn't hover on one engine, if we were going to get down in one piece, we needed to execute a running landing. To accomplish this, I was going to need a few hundred feet of unobstructed concrete, and they had about five miles of it over at Austin-Bergstrom International. We headed toward the airport, limping along on one engine, still flying at 70 knots.

During the short flight to Bergstrom, Stef did just what he had been trained to do and pulled out the emergency checklist we carried in the cockpit. He opened the book and, reading out loud, began listing the procedures for a "single-engine failure" until I stopped him.

"We're way the hell past that, Stef. Turn to the page that says 'single-engine landing.' And while you're at it, pull out the airport diagram for Bergstrom."

I knew that once our helicopter skidded to a stop, it wasn't going anywhere for a while, so I didn't see any reason to shut down an entire runway when a taxiway would serve our purpose without disrupting traffic into and out of the airport.

While I was on the radio, bargaining with the tower for a suitable landing spot, Stef was on another radio, making arrangements for a ground ambulance to transport our patient from the airport to the hospital. In addition, we were trying to keep our two fellow crew members, who were still treating the patient in the back of the helicopter, apprised of our situation.

Suffice it to say, we were a little busy during the short flight from North Austin Medical Center to the airport.

Bustin' a Gut

A few days after we had successfully landed our disabled helicopter on a parallel taxiway at Bergstrom International Airport, Stef Maier and I were called in to the front office for a debrief. At the time, we weren't all that concerned about it. In fact, we thought we might even receive some sort of commendation for the professional manner in which we had handled the emergency.

Instead, we received a lecture from our bosses on how we hadn't communicated clearly enough with the dispatcher after we'd aborted our approach to NAMC. It seems our bosses had been caught off-guard when they eventually learned we'd suffered an engine failure with a patient on board. Stef and I just looked at each other, incredulous that we were catching grief—all because we hadn't had time to paint a picture for the dispatcher while we were trying to deal with the small matter of an engine failure.

After a while, I could see Stef was just about to bust a gut, and so, wanting to protect my crew chief from saying something he might later have regretted, I stood up and began to speak. The office had a window, which offered a nice view out into the hangar bays. The EC-135 that we'd been flying on the day of the incident just happened to be parked in the bay closest to the office.

"You see that helicopter," I said, pointing through the window. "A few days ago, we were flying that thing at max gross weight—on one of the hottest days of the year—and we lost an engine on short final, the absolute worst time it could have happened to us. Take a good look at it!" I said, raising my voice a little louder. "There's not a scratch anywhere on it—so what the hell is the problem here!"?

To their credit, our bosses suddenly realized that the point I was making was not without merit. From that day forward, we never heard another critical word about how we'd handled the engine failure or the subsequent emergency landing.

Once they'd thought about it, I don't think our bosses wanted to see Stef Maier bust a gut any more than I did.

Posing next to the T-2C *Buckeye* after my first jet solo at NAS Kingsville, Texas, in 1984 (note the ejection seat is still intact, inside the cockpit).

View from my SH-60 *Seahawk* during a nice, tight formation flight in the Philippines (1987).

Sporting my Texas Rangers baseball cap on the flight deck of the USS *Valley Forge* (CG-50)—We were transiting the Strait of Malacca, on our way to Operation Earnest Will in the Persian Gulf.

Patrolling the sea lanes in the Persian Gulf (1987).

On short final to the USS *Valley Forge*—The cruiser's flight deck was a tight fit for our 21,000-pound *Seahawk*. In 2006, the *Valley Forge* was towed out to sea near Hawaii and used as a target ship. She now rests at the bottom of the Pacific.

The *Valley Forge* aviation detachment—That's me in the front row (2nd from left) with the "squared-away" haircut, kneeling next to Harold Graebe (far left).

Our return to HSL-43, at NAS North Island, following the six-month deployment to the Persian Gulf (1987)—That's yours truly, looking on whimsically as Harold Graebe makes a long-winded speech during the presentation of our cruise plaque to Commander George Galdorisi, our squadron CO. Galdorisi (hand on hip, in khakis) enjoyed a stellar Navy career, serving in several high-level command billets before retiring as a Captain. He is now a *New York Times* best-selling author, and I was honored when he graciously agreed to write the Foreword for this book.

Unloading a patient from the Bell 412 during the middle of the night (1993).

This is the EC-135 that Stef Maier and I were forced to land on one engine after experiencing an engine failure on short final. The EC-135 was the best-performing helicopter I ever piloted. The flight manual prohibited aerobatic maneuvers while flying it, but the sleek little rotorcraft was as graceful in a barrel roll as any *airplane* I ever flew. That is, if I had *actually* rolled it, which, of course, I would *never* have done. After all, it was a prohibited maneuver, and I was not one to operate outside of conventional boundaries. ... *Really*. Well, ... nobody can *prove* I rolled it, anyway.

13

The Longest Night

December 12, 2002

The warm Florida sun was setting behind us as we looked out over the Atlantic Ocean from the deck of a tiki bar on Jacksonville Beach. Jim Allday had volunteered to be the designated driver for the evening, but it was only a matter of time before he would also be burdened with the additional task of providing adult supervision for the rest of us.

We had spent the last three days at the Emergency Response Conference, an annual event sponsored by *Rotor & Wing* magazine. The magazine had flown the four of us to Jacksonville from Austin to receive the 2002 Helicopter Heroism Award as part of the convention, and this was our last night in town.

J.R. Esquivel had never been one to shy away from adult beverages, and this night was no different. Chris Jones-Piercy, the only female member of our party, was enjoying herself as well—responsibly of course. Being a connoisseur of superior lagers, I was dutifully sampling several of the local Florida brews in order to compare and contrast them with my favorites back in Texas.

Over the last few days, we had been paraded in front of people attending the conference as "heroes," and we were all a little tired of acting the part. Now we were ready to blow off some steam before returning to our day jobs back at STAR Flight.

As you may have already guessed, one thing led to another, and it wasn't long before the evening degenerated into a scene from the movie *Animal House*. A few hours after the sun had set, the party moved from the tiki bar to the beach across the street. Eventually, J.R. decided to go for a swim, whereupon he unceremoniously shed his clothes and disappeared into the surf. It was a moonless night, so we couldn't see him, but we could hear him splashing around and shouting something unintelligible about "riding a seahorse."

This went on for several minutes, and then things got even more interesting. Like Godzilla emerging from the ocean to destroy Tokyo, J.R. came stomping out of the water. When he hit the beach, he inexplicably broke into a full-on sprint, hightailing it past his clothes, . . . right past the three of us, . . . then off into the night. He was pickin' 'em up and puttin' 'em down, kicking up sand in his wake, as he headed straight for the tiki bar and the lights of downtown Jacksonville Beach.

Suddenly, I had a very real problem on my hands. J.R. was seconds away from being subject to arrest, and Jim, who was supposed to be the one keeping us out of trouble that night, just stood there with a puzzled expression on his face. Chris understandably showed no inclination to intervene, leaving yours truly with a difficult decision to make: Let a totally buck-naked J.R. cross the highway and get arrested . . . or run him down and tackle him—when he's totally buck-naked. Needless to say, I didn't have much time to fully consider my options. This had to be a split-second decision.

Opting for what I considered to be the lesser of two indignities, I quickly initiated my pursuit. Now, even on my best day, I couldn't outrun J.R.—but get him liquored up, and I might stand a fighting chance. Not being completely sober myself, I still wasn't confident I could catch him. Luckily, he stumbled a couple of times, and that was the break I needed.

Just a few yards from the highway, I dove and reached as far as I could. I barely clipped one of his ankles, but it was enough to bring him down. Wet and covered with sand, J.R. was still determined to make it across the highway. Fortunately, before my fleet-footed flight nurse (how's that for alliteration?) could scramble to his feet and resume his quest to spend the night in jail, Jim arrived on the scene and helped to subdue him. The two of us managed to pull him away from the highway, and after we were sure nobody could see us from a passing car, we went about convincing J.R. to get dressed.

The ensuing wrestling match seemed to stun J.R. and bring him to his senses, but the rest of us decided it might be a good idea to leave the beach anyway. It was well past midnight, and we'd been on a strict liquid diet since around four o'clock that afternoon, so we were all a little hungry. We decided to head back toward our hotel and, hopefully, find some place to stop for a late-night meal along the way.

This was a Sunday night (technically Monday morning), so our after-midnight choices turned out to be somewhat limited. With not many eateries available to us, we ended up stopping at a McDonald's. Jim parked the car, and we all got out to go inside. To our disappointment, the dining room entrance was locked, and we realized that only the drive-thru was open. We went back to the car so we could get in line at the drive-thru, but when we got in, we discovered J.R. was no longer with us. In light of his previous indiscretions on the beach, I was somewhat concerned that he might be getting himself into more trouble. As I got out and set about trying to ascertain J.R.'s whereabouts, I told Jim and Chris to stay in the car and get in line.

It didn't take me long to find him. The McDonald's was located in a shopping center, and J.R. had wandered over to a Walmart, where he'd commandeered one of their shopping carts from the parking lot. Stationing myself in the path of his misappropriated cart, I raised my hand and signaled for J.R. to stop.

"What are you doing, Junior?"

"Going shopping," J.R. said matter-of-factly.

With that, he executed a last-second evasive maneuver, narrowly missing yours truly as he continued rolling the cart toward whatever purpose he had divined in his head.

Wary that he might have dubious intentions, I kept my eye on him as he walked past me, toward the McDonald's. I turned, trailed behind him, and watched with concerned amusement as J.R. pushed the cart right up to the front of the McDonald's and began rolling it repeatedly into the locked door, trying to gain entrance into the dining room. Fortunately, I grabbed him and managed to pull him away before anyone else noticed the drunk guy trying to ram a shopping cart through the glass door. Forcibly relieving him of his battering ram, I sternly ordered J.R. back to the car while I returned the shopping cart to the Walmart.

J.R., who was always a man of his word, even when he'd had too much to drink, promised to go straight back to the car. Unfortunately, I hadn't made him promise to actually get *inside* the car. As I was walking back to the drive-thru, I could hear Jim and Chris screaming and laughing inside the car. There was J.R., squatting on the hood of the car—his exposed butt cheeks plastered, big as Dallas, against the windshield, directly in front of Jim's face (I believe I may have mentioned a few chapters back that J.R. was a bit of a liberty risk when he first came to STAR Flight).

Happily for all concerned, cooler heads eventually prevailed, and after a few tense and embarrassing moments, I was able to talk J.R. off the hood of the car and convince him to keep his pants pulled up.

In retrospect, I guess none of us were on our best behavior that night. It had been a little over a year since the events that had earned us the trip to the conference, and that was the first time the four of us had been together with a chance to unwind. J.R.'s overexposure notwithstanding, no one had gotten arrested, and after a hearty late-night serving of Big Macs and french fries, we'd managed to make it back to the hotel without further incident. It had been a long night, but aside from taking a hit to our reputations, everyone had survived, and no real damage had been done.

The next afternoon, while packing my suitcase for the trip home, I picked up the plaque that had been presented to me at the conference. As I placed it in the suitcase, I thought about my fellow STAR Flight crew members and the events that had occurred that night back in Austin, thirteen months earlier—a night that none of us would forget anytime soon.

The moment of truth, the sudden emergence of a new insight, is an act of intuition. . . . The diver vanishes at one end of the chain and comes up at the other end, guided by invisible links.

—Arthur Koestler (author)

In Austin, Texas, the 2001 winter solstice occurred on December 21, at 1:22 p.m. Central Standard Time. The night that began at official sunset that evening lasted thirteen hours and eleven minutes, making it the longest night of the year in the official astronomical records. For those of us at STAR Flight, however, the longest night of that year had already come and gone—a little more than a month earlier, on November 15.

The events leading up to one of the worst floods ever seen by the residents of Central Texas actually began thirty-six hours prior to that night, when a slow-moving low pressure system, packed with energy and moisture from the Gulf of Mexico, stalled north and west of Austin. The massive complex of storms camped there for the next two and a half days. The torrential rains it produced over a large portion of the state resulted in twelve deaths and more than a billion dollars in property damage, making it the deadliest and costliest Texas storm since the infamous Galveston hurricane in December of 1900.

Wednesday, November 14, approximately 11:45 a.m.

After coming off the night shift, a little more than thirty hours earlier, it was the first of three scheduled days off for the Leper

Colony. Stef Maier, J.R. Esquivel, and I had volunteered to fly the off-duty helicopter to Fredericksburg, seventy-five miles west of Austin, for a public relations event. I stopped by the clinical manager's office to chat with Jim Allday for a few minutes, then, at about the same time Stef and J.R. arrived, I headed out to the hangar bay to start preflighting our aircraft.

There was rain in the afternoon forecast, but nothing had developed yet, so we decided to go ahead with the PR flight and keep an eye on the weather. When we returned three hours later, the skies had begun to darken, but no rain had fallen. I wrote up a main rotor vibration in the aircraft maintenance log, and by the time we finished washing the helicopter and stuffing it into the hangar, some light rain had begun to fall.

At Brackenridge Hospital, Chris Jones-Piercy was the nurse assigned to the on-duty aircraft that day. She had just completed her crew chief training, and this was her first shift since her certification. As I was wrapping up my post-flight admin duties, I still had my radio on, and I heard the dispatcher assign Chris and her crew to a call in Llano, sixty miles to the northwest.

A few minutes later, just as Stef, J.R., and I were preparing to leave the hangar, we heard Chris go out over the radio to tell the dispatcher they were turning around for weather. By the time I was backing my truck into my garage an hour later, it was coming down in sheets. It continued to rain hard throughout the night and dumped more than eight inches of rain over most of the area. This was enough to saturate the ground and create significant runoff in the local creeks and rivers—but it was just a prelude of things to come.

Thursday, November 15, approximately 8:40 a.m.

When I woke up the next morning, the rain had stopped, but the skies still looked threatening. I had planned to mow my yard that day, but it was too wet, so I decided to work on my truck instead. The brand new, high performance throttle body I had ordered weeks ago had been delivered to my doorstep the previous day while I was out flying the PR event with J.R. and Stef. I spent the morning and part of the afternoon in my garage, removing the old throttle body and installing the new one.

I normally would have enjoyed a cold Shiner while I was working on my truck, but based on the forecast, I knew there was an outside possibility I might end up in the cockpit before the day was over. Even though I was scheduled to be off-duty for two more days, if it began to flood and we were forced to bring a second bird online, I would need to fly it at some point, so I elected to abstain and settled for a cold Diet Coke instead.

At around the same time I was opening my tool chest to start removing the throttle body from my truck, Spanky Handley and Mike Self (the on-duty mechanic) were wrapping up a track-and-balance flight on STAR Flight Two, the aircraft I had written up after returning from the PR mission the previous afternoon. Spanky and Casey Ping had already made the decision to put a second aircraft in service at the hangar.

Thursday, November 15, approximately 9:30 a.m.

Because fourteen inches of rain had fallen over Central and South Texas in the past fifteen hours, the Texas Division of Emergency Management activated Texas Task Force One, a federally coordinated urban search-and-rescue unit. The task force, made up of personnel from first-responder units throughout the state, met at Austin-Bergstrom International Airport that morning and began preparing for a deployment to Del Rio, on the Mexican border, which had already been hit hard by flooding. Among those who were headed to Del Rio was my flight nurse, J.R. Esquivel.

Thursday, November 15, 12:23 p.m.

The rain had been falling heavily again for an hour when STAR Flight One, the on-duty aircraft, was dispatched to search for a vehicle that had been swept from a low-water crossing near Blanco, fifty miles west-southwest of Austin. Chris Jones-Piercy, working her second consecutive day shift, was on board with Skip Gibbons, the pilot, and Stef Maier, who was working an overtime shift in place of the regular paramedic assigned to that crew. They searched for close to an hour, but they were unable to locate the vehicle. Finally, running low on fuel and faced with worsening weather conditions, they were forced to abort the search and return to Brackenridge Hospital.

Not long after returning to Brackenridge, STAR Flight One was dispatched to another water rescue, this time about thirty miles away, in Hays County, south of Austin. Because of the increasing intensity of the storm, they were forced to cancel the flight without launching.

Thursday, November 15, approximately 4:15 p.m.

As I was torquing the last bolt on my new throttle body, it began raining hard. A few minutes later, as I was placing the air cleaner on top of the newly installed throttle body, it suddenly sounded—and felt—as if my roof was being pelted with ball bearings. I looked out the window and saw water pooling in places where I'd never seen it before. The rain had literally begun coming down in a contiguous flow, as if my garage was at the base of Niagara Falls. As one of only two off-duty pilots, I knew it was time to call Casey Ping and ask him if he needed me to come to work.

"When can you get here?" Casey said, answering his cell phone.

"I'm on my way," I told him. "Just give me a few minutes to clean up and throw on a zoom bag."

As soon as I hung up the phone, it dawned on me that my wife, Nancy, had not come home from work yet, so my truck was the only vehicle I had at my disposal.

"I sure hope that new throttle body works," I muttered to myself as I went inside the house to round up my flight gear.

Thursday, November 15, 4:19 p.m.

STAR Flight One was dispatched to a water rescue for the third time that afternoon. Another vehicle had driven into a low-water crossing west of Spicewood, forty miles northwest of Austin. This time, however, the vehicle hadn't been swept away. Instead, with the driver trapped inside, it had become lodged in a tree.

By the time the STAR Flight One crew arrived over Spicewood, first responders on the scene had already managed to make their way into the tree and rescue the driver. STAR Flight One turned around to fly back to Austin, but the deteriorating weather forced them to land in a field next to State Highway 71, approximately thirty-five miles away from Brackenridge Hospital.

As they sat in the helicopter, waiting for the weather to lift, the crew was monitoring their hand-held radios. They heard the call go out to local ground units for a car wreck at a highway intersection not far from where they had landed. Stef Maier and Chris Jones-Piercy grabbed their medical gear and carried it, through the driving rain storm, from the field where they had landed to the scene of the wreck. Two of the accident victims needed transport to the hospital in Austin, so Stef Maier and Chris Jones-Piercy rode with them in the Spicewood ground ambulance while Skip Gibbons remained with the aircraft.

The weather improved as they drove southeast, and when they were still about twenty miles from Austin, STAR Flight Two, manned by Spanky Handley and Casey Ping rendezvoused with the ambulance.

After transferring the more critical patient into the helicopter, Chris Jones-Piercy, along with Casey Ping and Spanky Handley, flew him to the trauma center at Brackenridge Hospital. After delivering their patient, that crew returned to the Hangar in STAR Flight Two.

Stef Maier continued to Brackenridge with the remaining patient in the ground ambulance. After delivering the second patient to the hospital, the ground ambulance returned to Spicewood, and Stef Maier spent the rest of his scheduled shift alone at the STAR Flight crew quarters, where he waited for Skip Gibbons to return in STAR Flight One.

Thursday, November 15, approximately 4:25 p.m.

Much to my relief, my truck started right up when I turned the key in the ignition. It ran a little rough at first, but by the time I was about a mile down the road and on my way to the hangar, it smoothed out nicely. The rain had tapered off momentarily, but when I topped a hill on my usual route out of our neighborhood, the normally dry low-water crossing directly ahead of me was completely flooded with swift-moving, muddy water. I quickly turned around and took an alternate route out to the highway.

At around the same time I had been forced to alter my route to the hangar, a symposium on how to care for Alzheimer's patients was just about to conclude in Temple, seventy miles north of Austin.

As she was wrapping up the class she had been teaching as part of the symposium, Sharon Zambrzycki (pronounced *Zam-BRICK-ee*), an employee of the Texas Department of Human Services, received a phone call from her husband, who was calling to warn her about the severe weather that had descended on Austin. Sharon promised her concerned husband that she would be careful on her drive home to Pflugerville, just north of the city. It was still about thirty-five minutes until official sunset when she got into her car to start home, but that didn't matter much. The dark clouds and rain had already cast an early nightfall over Central Texas.

As Sharon made her way down Interstate 35, the rain became more and more intense. When she was close enough to receive a signal, she tuned her radio to a local Austin radio station, where she heard reports of numerous tornados and widespread flooding. After listening to the ominous reports, she made a conscious decision to take a route that would steer her clear of any low-water crossings. The final eight miles of the route she had chosen took her south along FM 685, a heavily traveled four-lane highway.

When she reached the bridge that crosses Brushy Creek at around 6:00 p.m., Sharon Zambrzycki was exactly 13.8 miles north of the STAR Flight hangar. She saw something in her headlights she had never seen there before that night—the water, normally twenty-five feet below the bridge, was moving across the highway.

Two Hours Earlier

In the face of the epic storm, the National Guard helicopter transporting J.R. Esquivel and the rest of the Task Force One team members was forced to turn around before reaching Del Rio. They had spent most of the day on the ground, west of San Antonio, waiting for a break in the weather. Now it would be getting dark by the time they landed back in Austin, and because Guard regulations prohibited rescue operations at night, they were essentially done for the day.

Thursday, November 15, approximately 6:00 p.m.

The sun had already set when I reached the hangar, and it was pitch-black outside. The rain was coming down in sheets again, and I got drenched just running the few yards from my truck to the door at the front of the building.

"Glad you made it," Casey said smiling.

"It was a lovely drive," I said sarcastically as I stood there dripping all over the floor.

"You're flying with CJP (Chris Jones-Piercy) and Allday," he told me as I went looking for a towel. "Chris is your crew chief."

Spanky Handley walked by at about that time and chuckled at my waterlogged condition.

"I don't think you'll be doin' much flyin' tonight," he said. "I've already had to cancel two rescues at Shoal Creek."

Shoal Creek runs through downtown Austin, not far from our hangar. *If we can't make it that far*, I thought to myself, *we're pretty much grounded.*

After telling me to "be careful out there," Spanky left the hangar through the same door I had used a few minutes earlier. He made a mad dash to his car, getting soaked in the rain just as I had. After picking up his wife from her nearby office, the two of them headed home through the storm.

After drying off, I made my way out to the hangar bay to start my preflight. Chris Jones-Piercy was standing next to the helicopter, along with Jim Allday, who had volunteered to stay and fly with me after working a full day in his office. Chris had been on duty since seven o'clock that morning, so even though I had started working on my truck eight hours ago, I was the freshest member of the crew.

While I was still preflighting the helicopter, we were dispatched to a "potential" rescue in western Travis County. A houseboat had been spotted floating down the Pedernales River, and the local first responders wanted us to determine if anyone was on board.

It would have been one thing to launch in that weather if we had been sure that someone was truly in imminent danger, but there was no way I was going to take that kind of risk just to determine if there was anyone aboard a floating houseboat. I told Chris, who was handling our radio traffic, to cancel the flight for weather.

Meanwhile, the rain was coming down even harder than when I had arrived at the hangar. The sound of the rain slamming into the hangar's exterior was almost deafening. It was also more than just a little unnerving as the huge hangar doors creaked and moaned under the stress from the wind.

I guess Spanky was right, I thought to myself as I continued preflighting. *It doesn't look like we're going anywhere tonight.*

Thursday, November 15, approximately 6:15 p.m.

When Sharon Zambrzycki realized the water moving across the Brushy Creek Bridge was surrounding her car, it was already too late to back up. In her rearview mirror, she could see headlights from another car directly behind hers, and cars were beginning to stack up behind that car as well. She dialed 911 on her cell phone, and the dispatcher told Sharon they had received a couple of earlier calls and that rescue units were already responding to the scene. By this time, Sharon could hear sirens in the distance, so she put her phone away and tried to maneuver her car, hoping to make it to higher ground.

Just then, Sharon saw another set of headlights coming toward her from the other side of the bridge. She could tell the water was rising rapidly as she watched the other car stall right next to hers—close enough, in fact, for Sharon to see the panic on the face of the woman behind the wheel. Because Sharon's car was still running, she rolled down her window and yelled at the woman to "get in!"

The noise from the rushing water made it difficult to hear, but apparently the woman understood that Sharon was offering to help her. She climbed out of her car to get into the back seat of

Sharon's car, but as the woman opened the door, the water suddenly began rushing inside, and she struggled to close the door behind her. Sharon knew the door hadn't closed completely because the dome light remained on after the woman got inside. The young man in the car parked directly behind Sharon's got out and waded up to her window.

"We gotta go! We gotta get out of here!" he yelled before heading back toward his car.

The young man got back inside his car, but the car wasn't moving.

Sharon hoped to maneuver her car out of the water, but now, in addition to the cars stacked in her rearview mirror, she was blocked by the car that had just stalled next to hers. She reached up to turn off the dome light so she could see outside, but before she could do anything else, the force of the water turned the car sideways. After a few more seconds, the woman in the back seat shouted at Sharon.

"The car's floating!"

The two women scrambled out of Sharon's car and into the rising water. Through the hammering rain, they tried to reach the car behind them, where the young man who had waded to Sharon's window was now standing, partially submerged, on the driver-side door sill.

The two women never made it to the man's car. A sudden, frigid, chest-high wall of water rushed across the highway and swept all three of them from the bridge.

This can't be happening, Sharon thought, trying to convince herself she would be okay. *This isn't real.*

That was the last time Sharon Zambrzycki saw the man and the woman who were with her on the bridge—all three went careening downstream and into the bitter-cold darkness.

Thursday, November 15, approximately 6:45 p.m.

Jim Spencer (pilot), Paul Kuper (crew chief), and Linda Jo Kuc (rescuer) arrived at Brackenridge Hospital to start their regularly scheduled shift, only to discover an empty helipad. Skip Gibbons was still stranded near Spicewood in STAR Flight One, unable to fly in the intense storm.

By this time, there were multiple tornados on the ground, but they were impossible to see in the heavy rain and darkness. The Austin/Travis County Emergency Operations Center was being overwhelmed with 911 calls to report people trapped in rising floodwaters all over Central Texas.

A little less than one hour earlier, J.R. Esquivel had touched down at Austin-Bergstrom International Airport, having returned from the aborted Task Force One deployment. His National Guard *Blackhawk* had made it back to Austin just minutes before the weather had become unflyable, shutting down the airport. Realizing he might be needed at STAR Flight, J.R. threw his gear into his SUV and headed to the hangar. It was rush hour and traffic was completely gridlocked in the storm. It would take him nearly fifty minutes to make the twelve-mile drive.

Thursday, November 15, 6:58 p.m.

I had just finished checking out the aircraft, when the pager went off for the second time since I'd arrived at the hangar. This time, the dispatcher told us there were three confirmed victims in the water at FM 685 and Brushy Creek. This was nothing like the *potential* rescue I had cancelled just minutes earlier—this was a no-kidding, people-are-about-to-die rescue. *Now*, I actually had a gut-wrenching decision to make—a life-and-death decision.

The weather was as bad as it had been all day. I looked out through the small, eye-level window in the hangar door, where, in the illumination from the high-intensity lights that surrounded our two helipads, I could see waves of water surging across the concrete. This was going to be the most agonizing go, no-go decision of my thirty-five-year flying career.

Casey Ping was no longer around. A few minutes earlier, he had left in his Suburban to go pick up Stef Maier, who was still stranded at Brackenridge Hospital. My crew and I were the only ones remaining at the hangar, which was okay by me. It gave me a welcome measure of solitude in which to make my decision.

A few years earlier in my career, I would not have had the confidence to fly in a storm of this magnitude. A few years later, after I had become wiser and less bulletproof, I would have cancelled this mission in a heartbeat.

For starters, I wasn't flying with my regular crew that night. This would be Chris's first rescue as a crew chief, and she and I had never even trained together, so I had no way to know how well she could actually *function* in that role. Not only that, she had already been on duty for eleven hours. As for Jim, he had worked a full day in his office. It was anyone's guess as to how fatigued *he* was by that point. As if that wasn't enough, because I hadn't been expecting to fly that night, I had spent most of the day turning wrenches on my truck.

Leaning my face toward the tiny window in the hangar door, my hands and elbows resting against the cold metal surface,

I could feel it shuddering from the wind and rain on the other side. As Jim and Chris patiently waited behind me, I closed my eyes and quietly gathered my thoughts for what must have been a full minute. It wasn't just my life on the line. There were two more people, standing right behind me, who were depending on me to make the right call. Spanky Handley, who was my boss, had turned down two rescues at Shoal Creek just an hour earlier, and I had already cancelled one for weather myself. Jim and Chris probably figured I would cancel this one as well.

Still leaning against the hangar door, eyes still closed, I lowered my head. I reached with my right hand and nervously rubbed my hair back and forth. Then I pulled forward and down against the back of my head until my chin was resting firmly against the top of my chest. I thought about the other people who were depending on me to make the right call—the three people in the water. They were seven minutes away from us, straight up FM 685, barely more than a takeoff and a landing from the hangar.

It felt as if I was trying to process a million thoughts all at once, but it ultimately came down to this: *Do I launch into the teeth of the storm and place my crew in harm's way, or do I play it safe and cancel?* One thing was absolutely certain. Nobody was ever going to question me if I decided to take a pass in this weather—nobody except for me, that is. No matter which way I chose, there was a chance I would end up regretting my decision.

My rational voice was shouting at me, telling me to turn the call down. My naval aviator voice was whispering in the background, telling me this was what I was trained to do. There was no getting around it, this was a moment that could define my career as a public-safety helicopter pilot—or possibly finish it.

Head still down, eyes still closed, I took a deep breath and forcefully exhaled. *Dammit!* I thought to myself. *I didn't come here to spend the whole night turning down rescue calls. If we can't go get these people when they're this close to our hangar, we should pack it up and go home.*

When I opened my eyes and turned around, Chris Jones-Piercy was holding her radio up, waiting to relay my answer to the dispatcher.

"Let's mount up!" I said.

Chris keyed her radio and said, "STAR Flight Two responding to FM 685 at Brushy Creek."

Jim Allday hit the button that opened the hangar door, and as it opened, we were instantly drenched by wind-driven masses of rain—blowing straight in toward us.

Thursday, November 15, 7:10 p.m.

As my ad hoc crew and I lifted from the helipad outside the STAR Flight hangar, Sharon Zambrzycki was in a desperate struggle to survive. After having been swept from the bridge, Sharon had been pulled beneath the surface of the water by her wet clothes. She had managed to kick off her shoes and slacks, and after successfully shedding the excess weight, she'd made it back to the surface, gasping for air in the darkness. Then, several hundred yards downstream, she had managed to grab a tree and hang on to it.

Now, as the water continued to rise around her, the intensity of the storm showed no signs of waning. The large pellets of cold rain, driven by the unrelenting force of the wind, were stinging her face and making it almost impossible for Sharon to keep her eyes open.

Sharon Zambrzycki was literally hanging on for dear life.

Soaked from having rolled the aircraft out of the hangar, Chris, Jim, and I were painstakingly making our way up FM 685, flying just above the streetlights. Because of the near-zero visibility, we were flying so slowly that we were getting passed by cars on the highway below us.

I remember thinking to myself, *Why are these idiots out here on the streets in this weather?*

In retrospect, I'm pretty sure there were people in those cars who wondered why there was an idiot flying a helicopter barely above the streetlights on a night when they were struggling just to navigate the highway.

Meanwhile, crabbing at more than forty-five degrees to the road in an effort to counter the gale-force wind, I was hoping I hadn't made a huge mistake. We were literally flying sideways through the storm.

At Brushy Creek, two Pflugerville firefighters, Tim Wallace and Trevor Stokes, were feverishly looking for survivors along the perilous banks of the raging torrent. The creek had risen nearly twenty feet, and it wasn't slowing down. Using their flashlights, they spotted Sharon Zambrzycki, still clinging to the tree she had managed to grab after being swept from the bridge. The two firefighters, neither of whom had ever participated in a swiftwater rescue, quickly anchored a rope on the bank. After rigging themselves to the other end of the rope, the two firefighters tried to reach Sharon, but the effort was futile. She was too far away.

They did manage to fight their way through the current to a tree—fifteen yards from the one where Sharon was frantically trying to hold on. Amazingly, Tim Wallace was able to toss her a life jacket and a rope. She managed to work her arms, one at a time, into the life jacket, and then she tied the rope around her waist. She managed to do all this while still holding on to the tree.

Thursday, November 15, 7:36 p.m.

After what seemed like an eternity, we arrived over a complex of soccer fields adjacent to FM 685, just south of Brushy Creek. Flying in near-zero visibility, we had been forced to creep along at barely more than 60 knots of indicated airspeed. Because we had been struggling to stay over the highway against the 40-knot crosswind, our actual groundspeed had been in the neighborhood of 35 miles per hour. What should have been a seven-minute flight had actually taken us twenty-six *excruciating* minutes.

There were two on-scene commanders on the ground—one from Williamson County, who was stationed on the north side of the creek, and one from Travis County, who was at one of the soccer fields on the south side. We were communicating by radio with the Travis County commander, who was interacting with us on one channel, monitoring the Pflugerville firefighters on another channel, and coordinating with the Williamson County commander on a third channel. This added even more confusion to an already chaotic situation. Because of conflicting reports being passed to the two commanders on opposite sides of the creek, no one could agree on where we should begin searching for the three people who had been swept from the bridge.

We had already expended too much time just trying to get here. We were determined to start looking somewhere, even if it wasn't where they wanted us to concentrate our search. We marked on top of the bridge and headed downstream, sweeping the water and trees with our night sun.

Within minutes, we had spotted one victim, three hundred yards from the bridge.

Thursday, November 15, approximately 7:40 p.m.

As she desperately clung to the tree where she was stranded, Sharon Zambrzycki struggled to keep her head above the water. Suddenly, a bright light was overhead. She couldn't hear it over the deafening roar from the water rushing by her, but she knew it was a helicopter. Sharon's hopes were buoyed until, incredibly, it began raining even harder. Then, just as quickly as it had appeared, the light flew away. Sharon was devastated. She didn't know why the helicopter had flown away. Surely they had seen her.

Then it happened. The current, moving at more than 40 miles per hour, snapped the branch and sent Sharon barreling downstream again. This time, the limb that had been her refuge a few seconds earlier became tangled in the rope around her waist. Instead of floating on top of the current, the limb entered a standing wave and pulled her under the dark, murky water.

No longer able to see and sinking deeper below the surface, Sharon didn't know how long she could hold her breath. She struggled to free the broken limb from the rope around her waist, but it was hopeless. Then the huge limb exited the eddy beneath the standing wave and propelled Sharon to the surface, where she slammed against the trunk of another tree. The tree was too big for Sharon to get her arms around it, but she did her best to hold on and keep her head above the water.

She knew there was nothing more the two firefighters could do for her at this point. Sharon Zambrzycki had been in the deadly water for more than an hour now. Shivering from the extreme cold, she was battered, exhausted, and just about ready to give up hope.

Just then, the floodgates opened even further, and an incredible tsunami of debris-filled water came roaring down the already-out-of-control creek. The surge came upon them so quickly that Tim Wallace and Trevor Stokes were now trapped in the flood as well.

Wallace began yelling at Sharon to "hang on!"

"Keep your head above the water!" he shouted. "We're going to get you!"

Sharon strained to climb up the tree, but the rope around her waist was holding her down. She frantically tried to free herself from it, but she couldn't. Wallace kept shouting at her to keep her head out of the water.

"I'm trying!" she shouted back.

The water continued to rise around her, and she had no way to move higher into the tree. She closed her eyes as the rising water began to pound her face. Sharon knew she was about to drown. She remembered reading somewhere that drowning was a peaceful experience.

Then she thought to herself, *How do they know that?*

Instead of surrendering, she began to get angry, refusing to accept her seemingly inevitable fate. She thought about her husband.

He's going to be pissed. I can't leave him.

Then she thought about her children. She tried—desperately tried—one last time, to move up the trunk of the tree.

Thursday, November 15, 7:48 p.m.

At about the same time Sharon Zambrzycki was being swept from the tree where we had just spotted her, we landed on the soccer field next to the command post, and Chris Jones-Piercy rigged the short-haul as quickly as I'd ever seen it done. Jim Allday ran over to say something to the on-scene commander, and by the time he returned, Chris was meeting him at the front of the helicopter to make sure he fastened himself properly to the line. A few seconds later, we were lifting Jim on the short-haul.

We flew back to the tree where we had spotted the victim, and that's when we realized she was no longer there.

"She's gone!" Chris yelled over the wind and driving rain.

We were both getting soaked from the deluge coming in through the open cargo door, and I worried we might short out the ICS, along with some of the other avionics in the cockpit. Frustrated at having lost her, we started moving downstream with Jim still on the short-haul. Several minutes later, we once again spotted the victim through the driving rain, in a different tree. We could tell she was struggling just to keep her head above the water.

"Let's not waste any time," I told Chris.

Because Chris was a rookie crew chief, and because we had never trained together, I was still apprehensive about how this was going to unfold. This wasn't just her first rescue as a crew chief, it was only her second shift since qualifying.

Without missing a beat, Chris began guiding me into position as if she'd been doing it her entire life. If it's possible to yell and be calm at the same time, Chris was doing it that night.

Thursday, November 15, approximately 7:55 p.m.

Her eyes still closed, Sharon Zambrzycki was just about to slip beneath the surface of the water when, suddenly and unexpectedly, she sensed thirty million candles of light—the beam from our night sun—shining on her face. Holding her breath, she opened her eyes just in time to see Jim Allday, suspended on the short-haul line, descending directly in front of her. Even though we were fighting gale-force winds and blinding rain, Chris had placed Jim in the perfect spot to get his arms around Sharon. He quickly got the rescue ring around her and signaled to Chris that he was ready to come up.

As they began moving, Sharon remembered the rope around her waist. Out of breath and unable to make herself heard over the noise from the helicopter and the roar of the rushing water, she tried pointing toward the rope. Fortunately, both Jim and Chris had already spotted it.

"Hold your hover!" Chris yelled to me. "Down three!"

Jim took out his knife, and with several deliberate but rapid strokes, he slashed the rope, freeing Sharon from her deadly tether.

"Easy up!" Chris shouted, once again making sure I could hear her above the noise of the storm.

Thursday, November 15, 7:59 p.m.

Nearly two hours after she had been swept from the Brushy Creek Bridge, Sharon Zambrzycki's feet gently touched down on the muddy soccer field. Several paramedics from the command post were already moving toward her with a backboard and some blankets. She didn't know it yet, but her body temperature was at 92.7 degrees, bordering on severe hypothermia.

"I can walk!" she yelled to Jim Allday, with the helicopter still hovering overhead.

Jim took the rescue ring from around her, and she took three steps before falling to the ground. The paramedics placed her on the backboard and covered her in blankets before loading her into an ambulance.

Thursday, November 15, 8:02 p.m.

Jim Allday gave us a thumbs-up and we returned for Tim Wallace and Trevor Stokes, the two firefighters who had risked their lives to save Sharon Zambrzycki, only to become trapped in the flood themselves. Chris continued to be spot-on with her crew-chiefing skills, and because of their training, rescuing the two firefighters turned out to be as routine as rescues can be during a hundred-year storm.

We then turned our attention to the two remaining people who had been swept downstream from the bridge. We continued searching for another forty minutes, battling the wind and rain, until we had to bingo back to the hangar. The search would have to wait until we could refuel. I hoped the flight back to the hangar wasn't going to take another twenty-six minutes.

Thursday, November 15, 9:17 p.m.

When we landed back at the hangar, J.R. Esquivel was there to meet us. The rain had started to slacken a little during our return flight from Brushy Creek, but the damage was already done. A little over fifteen inches of rain had fallen in the past two hours, and the runoff into the rivers and creeks was going to keep us busy for the rest of the night.

Immediately after I had secured the engines, J.R. began refueling the helicopter. The blades were still coasting to a stop when he yelled to ask us if there was anything else we needed.

"Did you bring doughnuts!"? I shouted back.

I was only half joking. It had been ten hours since I'd had anything to eat, and I knew there wasn't any food at the hangar.

"Are we going to shove it back inside?" J.R. asked.

"Negative," I answered. "We have to head back to Brushy Creek. We still have two people missing out there."

Chris Jones-Piercy had been on duty for more than fourteen hours, so J.R told her to go home, he would take her place. Of course, J.R. had already put in a full day with Texas Task Force One, but he wasn't about to stand on the sidelines now. J.R. was as strong as an ox and stubborn like a mule. He was going to jump into the fight and stay until the battle was over, and it really didn't matter how long he'd been on the clock or what anybody else thought about it.

Chris gathered up her gear and began walking through the rain, toward the hangar. I yelled at her, and she turned around, still walking away.

"You can crew chief my bird anytime!" I shouted at her.

Chris smiled, waved goodbye, and turned back around.

Forty-Five Minutes Earlier

At about the same time we were rescuing the two Pflugerville firefighters at Brushy Creek, Skip Gibbons, stranded in STAR Flight One, had finally gotten airborne again and made it back to Brackenridge Hospital. There, Jim Spencer, Paul Kuper, and Linda Jo Kuc were waiting to begin their night shift. Skip turned his aircraft over to Jim Spencer, and while Jim was preflighting, Skip dragged the fuel hose over to refuel it.

When Skip turned on the pump to begin refueling the helicopter, nothing happened. The storm had knocked out the hospital fuel system, and there wasn't enough fuel remaining in the aircraft for Jim Spencer and his crew to fly it to the hangar.

Casey Ping instructed Linda Jo Kuc to drive to a nearby hardware store to purchase some fuel cans. She made it to the store just before it closed, and after purchasing six five-gallon cans, Linda Jo headed to the hangar, where she arrived at just about the same time my crew and I were returning from Brushy Creek.

She filled the cans with thirty gallons of fuel and returned to Brackenridge, where she and the rest of her crew emptied the cans into STAR Flight One. It was enough to get them to the hangar, where they would then be able take on a standard fuel load.

Thursday, November 15, 9:52 p.m.

When Jim Spencer and his STAR Flight One crew landed at the hangar, on the pad next to ours, J.R. Esquivel, Jim Allday, and I were already preparing to head back to Brushy Creek to resume our search for the two people who were still missing. Just as I was about to light the first engine, we were reassigned to a new rescue, this time in South Austin, on the other side of town. As soon as we launched and made our turn to the south, it started raining hard again.

A few minutes later, after they had refueled and reported themselves in service, the crew of STAR Flight One was dispatched to resume searching for the two people still missing in Brushy Creek.

The night was still young.

Thursday, November 15, approximately 10:10 p.m.

Colton-Bluff Springs Road is located about one mile southeast of Austin-Bergstrom International Airport, which was now reopened. The airport was still IFR because of the low ceiling and poor visibility, so we had to get a "special VFR" clearance from the tower. We were trying to reach a motorist who had become trapped in the middle of the flooded roadway. When we arrived, it appeared the man had literally smashed his way through the windshield to escape the car, and now he was stranded on the hood.

Following Chris Jones-Piercy's departure, Jim Allday had assumed the role of crew chief, and J.R. Esquivel was now my rescuer. After finding a place to set down and rig the short-haul, we lifted off again and lowered J.R. onto the hood of the car. To avoid slipping off the wet hood, J.R. was forced to brace himself against the broken windshield and sliced his hand on the jagged glass. Ignoring the injury, J.R. managed to get the rescue ring around the man and signaled to Jim that he was good to go.

It continued to rain, and as we lifted them both from the hood of the car, the wind started blowing hard once again—almost as hard as it had been blowing during Sharon Zambrzycki's rescue. I was getting tired of trying to hold a steady hover in a hurricane, so after setting them down and recovering J.R., I was hoping to make a hasty retreat to the hangar. But before we could do that, the on-scene commander came over the radio and informed us of another missing motorist on the same road.

We searched for about twenty minutes, but while we were still in the process of looking for the other vehicle, our dispatcher once again diverted us to a different rescue. This one was already in progress, about a mile and a half away.

Thursday, November 15, 10:28 p.m.

During their attempt to locate the remaining two people who were missing in Brushy Creek, the crew of STAR Flight One was diverted to still another rescue in Pflugerville, not far from where they were searching the flooded creek. When they arrived on scene, in an inundated rural neighborhood, they found four members of a family trapped on the roof of their home.

With Paul Kuper crew chiefing and Linda Jo Kuc on the short-haul, they rescued the family members, one by one, and then returned to Brushy Creek. They searched for another hour before returning to the hangar for fuel. The two people who, more than three hours earlier, had been washed from the bridge with Sharon Zambrzycki were still missing in the flood.

The next afternoon, searchers found the body of the woman who had taken refuge in Sharon Zambrzycki's car approximately six hundred yards downstream from where we had rescued Sharon. A few hours later, the body of the man who had been driving the car behind Sharon's was also discovered—more than six miles away.

Thursday, November 15, approximately 11:10 p.m.

When my crew and I arrived over the rescue operation to which we had been diverted just minutes earlier, it was already a beehive of activity. The on-scene commander advised us that two Austin police officers, while investigating a report of someone missing in Onion Creek, had become trapped in the flood themselves. They had been searching downstream when they were hit with a swell of water that forced them both into the trees along the creek. Now there were additional officers trying to reach them from the bank, and there was also a boat crew from Austin EMS. The boat crew had just arrived, and they were preparing to enter the water.

We spotted one of the stranded APD officers with our night sun and advised the on-scene commander that we were going to land so we could deploy our rescuer on the short-haul. He told us to stand by. He wanted us to hold our hover and keep the light on the two officers while the Austin EMS crew attempted to rescue the two men by boat. We watched as the three crew members launched their outboard-powered zodiac into the raging water beneath us.

I told my crew, "Get ready, guys. We're about to have five people to rescue instead of two."

I had no sooner said it, when the boat was swamped, and the three paramedics were swept into the trees, not far from the two APD officers. Without waiting for any more instructions from the on-scene commander, we landed and put J.R. on the short-haul line. Shortly thereafter, we were airborne again, trying to get the first APD officer out of the tree where we'd spotted him minutes earlier.

Unfortunately, the synergy we had enjoyed when Chris Jones-Piercy was crew chiefing earlier that night was no longer there. Jim Allday and I were both getting fatigued by this time, and it was starting to show. Jim tried his best to guide me in, but we kept dunking J.R. under the rapidly moving current that was ripping through the trees. Every time Jim told me he'd lost sight of J.R., I would come up on the collective to pull him from under the water. In doing so, we kept raking J.R. through the tree limbs. In spite of the difficulty Jim and I had coordinating with one another, we managed to short-haul the two police officers to safety, but poor J.R., who had already injured his hand on the previous rescue, was taking a beating in the process.

When we pulled the first of the three Austin EMS paramedics out of the trees, he happened to be someone with whom my crew and I were very familiar. Although Landon Wilhoite was no longer flying with us by then, he had been one of the original cadre of STAR Flight crew-chief instructors. Prior to that, he had been a PJ (Pararescueman) in the United States Air Force, which is the Air Force equivalent to a Navy SEAL. Landon was as tough as they come, and before becoming an Austin EMS paramedic, he had already survived several helicopter crashes while serving on active duty.

From his vantage point in the trees, Landon Wilhoite had watched us bang J.R. Esquivel around like a crash dummy while rescuing the two police officers. He knew J.R. had to be tired, so after J.R. had pulled him from the trees, Landon asked to take a turn on the short-haul line so that J.R. could rest. Not surprisingly, J.R. maintained that he was fine, but Landon insisted he needed to take a break, and Landon Wilhoite was not a man with whom you wanted to quarrel.

J.R. was as persistent as Landon was strong, but he knew he couldn't waste any more time arguing about it. Reluctantly, he surrendered the rescue harness to Landon and watched the rest of the evolution from the ground. When we lifted again, this time with Landon on the short-haul, Jim and I finally began to sync up a little better.

After Chris Jones-Piercy's performance earlier that night, I had begun to think that maybe crew chiefing wasn't that hard after all. I wondered if I'd been giving Stef Maier more credit than he actually deserved. Now I decided Stef Maier deserved all the credit I had been giving him—and more.

It wasn't as if Jim Allday was a rookie. He was "Mr. STAR Flight." He was respected by his supervisors and fellow crew members alike, and he was exceptional at just about everything he did. If he was struggling that night, it only demonstrated how tough it actually was to function as a crew chief—especially after you'd been on duty for eighteen hours.

There's almost a Zen-like aspect to crew chiefing a rescue, and the truth of the matter is, there just weren't that many people who were highly skilled at it. Stef Maier was the exception. Would I have been more comfortable with my regular crew chief in the back of my bird just then? Probably so. Was I happy to have my longtime, trusted colleague there in his stead? Absolutely. As far as I was concerned, Jim Allday, who had to have been even more exhausted than I was by that time, was doing a commendable job under the toughest imaginable conditions.

Friday, November 16, approximately 12:05 a.m.

As Landon Wilhoite was pulling the last of his fellow paramedics from the trees, I heard Jim Spencer, in STAR Flight One, asking for a special VFR clearance from Bergstrom Tower.

According to FAA regulations, only one special VFR aircraft at a time can fly within an airport's air space, and we were already operating on a special VFR clearance of our own. At first, the tower controller hesitated; but Jim explained that they were on a rescue mission, and the controller cleared STAR Flight One into the airspace not far from where we were working.

Once they had received their clearance, Jim Spencer, Paul Kuper and Linda Jo Kuc rescued two more people from another flooded neighborhood, pulling one from a tree and the other from yet another rooftop.

Having completed those two rescues, they were able to make their way back to the STAR Flight hangar just before the weather became completely unflyable again.

Friday, November 16, 12:26 a.m.

Having successfully completed all five short-hauls from Onion Creek, we landed to recover J.R. for the flight back to the hangar. By this time, we were running low on fuel and the ceiling was coming down fast. Casey Ping let us know, by radio, that the Brackenridge Hospital system was back in service; so I flew west for about a mile and then tried to fly north, up Interstate 35, hoping to make it to Brackenridge.

Along the way, the ceiling kept coming down lower and lower. When we reached a tall flyover intersection, where U.S.

Highway 290 merges onto Interstate 35, the cloud deck was too low for me to climb over the elevated ramp.

I was tired, hungry, and not in the mood to risk going inadvertent IFR at that point, so I turned around and looked for a place to set it down.

Rest for the Weary—Doughnuts for the Hungry

As luck would have it, the first place I could find to land was a vacant lot next to a Krispy Kreme doughnut shop, about a mile south of where we had turned around.

"This works for me," Jim Allday said happily as we were on short final.

We spent the next two hours eating doughnuts and waiting for the weather to lift. While we were there, the Austin/Travis County EMS medical director, who had heard us radio our dispatcher that we were making a precautionary emergency landing, stopped by to check on our status. He sat down at our table to drink a cup of coffee and chat.

"How are you guys holding up?" he asked.

"We're fine," I told him facetiously. "We were just flying along, and then we saw this Krispy Kreme sign and couldn't pass it up. We decided to pull over and take a short break, but as soon as we finish eating, we'll get back out there in the fray. We haven't even broken a sweat yet."

Truth be told, though, all of us were dog tired. I certainly was. The concentration required to repeatedly hover for extended periods while trying to follow the crew chief's verbal commands is extremely fatiguing to a pilot—even in good weather. When you're getting battered by wind and driving rain, it'll wear you down in a hurry. It had been a long night, and I was ready for it to end.

At around three o'clock in the morning, the ceiling had finally lifted enough for us to launch. The weather was still marginal, but we were able to clear the tall flyover at the highway intersection and make our way back to the hangar. When we landed, Willy Culberson, the man who had liberated me from my exile in the chief pilot's office, was waiting there to relieve me.

I can't tell you how happy I was to see Willy there. I had been going at it for nearly twenty straight hours, as had J.R. Esquivel. I had mostly been beaten up mentally, but J.R. had actually been beaten up physically. Even after all we'd been through, however, J.R. still wasn't ready to throw in the towel—which is a good thing because there wasn't anyone there to relieve him anyway.

And as for Jim Allday? Well, he ended up having the longest night of all. He and J.R. Esquivel would launch two more times that night, flying with Willy Culberson. They wound up rescuing five more people, and when the night finally ended, STAR Flight crews had rescued a total of twenty people from the deadly floodwaters. When it was finally over, Jim Allday, Mr. STAR Flight himself, had participated in fourteen of those rescues, more than any other STAR Flight crew member.

All things considered, it hadn't been a bad night's work.

LIFE INSIDE THE DEAD MAN'S CURVE

In the days following her rescue from Brushy Creek, Sharon Zambrzycki would make a full recovery. She had fought hard to stay alive that night—harder than anyone can imagine. Battered, cold, and exhausted, she could have easily given up during her furious struggle to survive. In persevering against the odds, she had displayed an abundance of courage and physical tenacity that belied her diminutive stature. It would seem implausible that such a large measure of willpower and fortitude could be contained in such a small frame.

Sharon has often thanked my crew and me for saving her life, but it was only because she steadfastly refused to surrender to the violent flood that we were able to reach her in time to keep her from drowning. When faced with her decisive moment of truth, Sharon's "sudden emergence of a new insight," as Arthur Koestler described it, was the simple realization that she wanted to see her family again. Sharon Zambrzycki had vanished at one end of Koestler's chain and then reemerged at the other end—where she discovered Jim Allday waiting for her. It was her will to survive that had provided the invisible links.

I want to find that defining moment when you're satisfied, and you've done what you want to do in your life.

—Tyson Gay (athlete)

In the days and weeks that followed, there were plenty of Monday-morning quarterbacks on the EMS helicopter blogs who said we'd had no business flying on the night of the floods—especially during the first three rescues at Brushy Creek. Some even said it was only a matter of luck that we hadn't added ourselves to the long list of EMS helicopter crews who have perished after exercising poor judgment.

They were probably right, of course. I definitely know that, later in my career, I would never have launched into conditions as severe as those we'd endured during the twenty-six-minute flight to reach Sharon Zambrzycki. In spite of that, as I write this—and I have to be honest here and admit that I have the luxury of 20/20 hindsight—I'm still comfortable with my decision to challenge the storm that night, even in the wake of criticism from those who said it had been the reckless act of a cowboy. Maybe we *were* lucky, but I prefer to think my fellow crew members and I rose to the occasion during those rescues. For me personally, the decision to launch on November 15, 2001, turned out to be the defining moment in my tenure as a STAR Flight pilot.

When I finally got home, just before sunrise that next morning, I was exhausted and hungry. All I had eaten in the last twenty hours were those Krispy Kreme doughnuts we'd consumed while waiting out the weather. Even so, I headed straight upstairs to the bedroom without stopping to eat breakfast. Stripping off my flight suit, I hit the rack and immediately went to sleep.

For ten straight hours, I slept.

I slept better than I had ever slept in my life—better than I'm likely to ever sleep again.

Storm clouds gather ominously over the STAR Flight hangar.

Central Texas is notorious for deadly flash-flooding events.

A STAR Flight rescue swimmer being hoisted from the water.

Sharon Zambrzycki in 2001, several months after her rescue from Brushy Creek during one of the worst storms to ever hit Central Texas.

Sharon meets with me and my ad hoc crew, which was pieced together on the night of the storm—(left to right) yours truly, Jim Allday, and Chris Jones-Piercy.

In November of 2002, STAR Flight made the cover of *Rotor & Wing* magazine.

Standing in front of the Texas state capitol is STAR Flight's award-winning crew (L-R): Jim Allday, clinical manager; Chris Jones-Piercy, flight nurse; Kevin McDonald, pilot; and J.R. Esquivel, flight nurse.

The magazine took this photo of my crew and me several weeks before flying us to Jacksonville, Florida, to receive the 2002 Helicopter Heroism Award. The photo was published with the article that documented our rescues during the 2001 flood—(left to right) Jim Allday, Chris Jones-Piercy, yours truly, and J.R. Esquivel (who relieved Chris halfway into the night).

14

Two Weeks in September

Nancy and the kids could always tell what I'd been doing when I came home after firefighting missions. The smell of smoke on my flight suit was a dead giveaway. In 2011, for a fifteen-day period that began in late summer, the smell of smoke was everywhere in Central Texas. Fueled by vegetation dried from several years of severe drought, fires were lighting off all across the area. There was literally no place inside a seven-county area where you couldn't see a wildfire somewhere on the horizon.

    I'm not exaggerating when I say that, for the crews at Travis County STAR Flight, there was a war looming in our futures, and it was a war we were destined to lose—a war against Mother Nature herself.

Lacking the resources to win the war outright, the only viable option we had was to engage in a two-week-long tactical retreat. Hoping to minimize the casualties, we waged a hit-and-run campaign while we waited for the advancing attacker to run out of momentum. The first major battle took place on the 4th of September. That was the day I witnessed the chaos that results when ten thousand residents are forced to evacuate a community that has only one road leading to safety.

Steiner Ranch is an upscale housing development located about twenty miles west of Austin, near Lake Travis. My house was ten miles away, but the wall of smoke from the fire that was rapidly consuming homes was clearly visible from my backyard. Late in the afternoon, as I stood there watching it, my cell phone rang. I took a quick look at the caller I.D., but I needn't have. I already knew who it was. Even before I answered it, I knew I was about to be pressed into service. I had just come off a night shift that morning, and I knew Casey Ping wouldn't have been calling me unless he was out of pilots.

"Can you fly?" Casey said apologetically.

"I can be there in forty minutes," I answered.

"Stay where you are," he said. "We'll come get you."

A few minutes later, I was waiting at the Lakeway Airpark, about a quarter mile from my house, when one of our three EC-145 helicopters, piloted by Kenny Thompson, a former Army pilot, landed to pick me up. I jumped into the back, and we immediately took off, headed toward the flames beneath the wall of smoke just a few miles away.

It looked like a scene from a Hollywood disaster movie, only this was real. I couldn't tell how many homes were actually on fire because of the thick, black smoke obscuring the ground. What I *could* see was a line of cars that stretched for miles as people were trying to evacuate. It looked like we had flown into the apocalypse, as the never-ending trail of headlights, along with the flashing lights from the army of fire trucks, was barely visible against the backdrop of the devastating smoke and flames.

We landed in a church parking lot, which was now a staging area, across the highway from the entrance to the fire-engulfed community. I hopped out and watched as Kenny, along with Patrick Phillips (his crew chief), took off and resumed fighting the fire. STAR Flight had commandeered the parking lot and converted it into a temporary heliport, complete with a refueling trailer. The rolling gas station was manned by my old confidante, Mike Self, who had driven it from our hangar, thirty miles away. It had been over a decade since I had routinely hidden out in Mike's office during my chief pilot days. I was always happy to visit with him, but neither one of us had the time for casual conversation just then.

From my vantage point at the church parking lot, I could see Kenny's bird, along with another STAR Flight helicopter, making repeated drops on the fire. The second helicopter was being flown by my longtime friend, Jim Spencer, a former naval aviator and onetime squadron mate of mine.

There was still about an hour of daylight left, and I was anxiously waiting, along with Kristin McLain, to join the fight. Called in to fly as my crew chief, Kristin had been scheduled to be off duty that day, just as I had; but with much of the county being threatened by wildfires, it was all hands on deck. The third STAR Flight helicopter was being flown by the man who was the current chief pilot, Mark Parcell (also a former naval aviator) and Lynn Burttschell, his crew chief. They were busy fighting another wildfire, farther west, near the Pedernales River.

Kristin and I were joined at the temporary heliport by Scott White. "Scooter," as he was better known, had been a Forward Air Controller in the U.S. Air Force, and while serving in that role, he had often been called upon to coordinate combat air strikes from a position in close proximity to a target. Now, as a STAR Flight crew chief, he was part of a different kind of aerial assault, a desperate attempt to save a community from total destruction by fire. Scooter was waiting for an aircraft, just as Kristin and I were. It wouldn't be long until the crews who were currently fighting the fire would need to be relieved, and we were standing by to take their places.

Scooter would be flying with the other relief pilot, Chuck Spangler. Like Kenny Thompson, Chuck had earned his wings in the Army. After a long career in the National Guard, he had retired as a lieutenant colonel before hiring on with STAR Flight. During his time as a STAR Flight pilot, Chuck had become one of my closest friends, and in addition to being the best amateur golfer I had ever seen, he was a pretty decent aviator as well. Fortunately for everyone who was depending on us that day, I was much more reliable in the cockpit than I was on the golf course.

Just before 7:00 p.m., Jim Spencer, who had been flying all day and was running low on fuel, landed to turn his aircraft over to me. His crew chief was J.R. Esquivel, my old flight nurse from our Leper Colony days, and both Jim and J.R. looked as if they'd been "rode hard and put away wet," as we say in Texas.

"It's like pissin' in the wind," Jim said as he climbed down from the cockpit. "There are too many houses burnin' and too few of us to do anything about it."

Kristin and I quickly checked out the aircraft while Mike Self was refueling it. I climbed into the cockpit just as Mike was removing the fuel nozzle, and as soon as he gave me the signal that we were good to go, I started lighting the engines.

After running the aircraft up and going through my takeoff checks, I made sure Kristin was ready, and then I raised the collective. We lifted our bucket, which was hanging about 15 feet below us, straight up from the parking lot. Once Kristin had told me the bucket was clear of the power lines, I lowered the nose and transitioned to forward flight, accelerating past the long line of escaping cars on the highway next to us. We headed straight toward Mansfield Dam on Lake Travis, where we filled our bucket and made our first pass at the fire.

At about that same time, Mark Parcell and Lynn Burttschell returned from the Pedernales fire to refuel at the staging area. They turned their aircraft over to Chuck Spangler and Scooter White. Because it was so close to sunset, Chuck and Scooter, instead of flying to the Pedernales fire, stayed to help fight the Steiner Ranch fire.

Now we had all three of our birds dropping water at Steiner Ranch, but it still wasn't nearly enough. Even with dozens of firefighters on the ground, we were facing an enormously daunting task. As quickly as we threw water at one house, two more would light off. As we continued into the night, it was about to become even more challenging.

"What are you doing up there?"

The sun was setting behind the hills west of the lake as all three STAR Flight helicopters were hovering in the shadow of Mansfield Dam. At 10 feet above the water, our rotor wash was misting our windscreens as we filled our buckets. Kenny Thompson was making drops to the north, along the leading edge of the fire, which was where firefighters on the ground were desperately bidding to make a stand against the advancing conflagration.

Chuck Spangler and I were dropping on the back side of the fire, struggling to save three houses that bounded a cul-de-sac. Situated at the bottom of a hill, the houses were the only ones left at the end of a street where nearly twenty were fully engulfed in flames. Chuck and I each made four drops on the houses immediately uphill and adjacent to the three still standing before we determined that the tactic wasn't going to work. It was obvious that trying to extinguish the dozens of homes that were already ablaze was useless. The fire was burning too hot, and much of the water was evaporating on the way down, so we decided to start making drops directly onto the houses we were trying to save.

Normally, we would have dropped only on structures that were already lost to the fire, but this was anything *but* a normal situation. We might do some structural damage attacking the homes directly, but if we didn't get some water on them quickly, they were going to be consumed by the fire anyway, just like the other homes on that street. Jim Spencer had already said it—there were too many houses on fire, and we didn't have nearly enough resources to save them. There was a slim chance, however, that if we concentrated our drops on the three that weren't yet burning, we just might be able to save them.

Chuck and I were flying a daisy chain as we began attacking the houses directly, timing it so that one of us was dropping while the other was filling his bucket. We were evenly staggered, which gave each of us time to locate the other's bird on the half-mile flight between the lake and the drop zone. We had made close to ten of these circuits when it started getting dark.

Although nighttime firefighting missions were not encouraged, the official STAR Flight policy stated that a pilot had the option to continue fighting a fire at night, provided he had begun fighting it during daylight. The rationale for the policy made sense. If a pilot could figure out where the obstacles around the fire were located during daylight, he could take the appropriate measures to avoid them once it was dark.

There was certainly no shortage of obstacles surrounding our cul-de-sac. Originating at the hydro-electric generating station beneath Mansfield Dam, there were two sets of high-tension power lines in our path. Both ran straight up the hill, one set on each side of the street we were working. Standing 50 feet in height, one set was on our approach to the cul-de-sac, and the other was directly in front of us on our climbout. Knowing they were there was just part of the equation. Getting in and out without hitting them or dragging our buckets through them was the real challenge.

I called Chuck Spangler on the radio to ask him what he thought about continuing after dark. Although we hadn't yet used them on firefighting missions, we had been equipped with NVGs (night-vision goggles) for quite some time. We had routinely flown with goggles on every other type of mission since 2007, four years earlier. They were the latest, most up-to-date NVGs money could buy, and we had used them with great success—just never while fighting a fire. Of course, we'd never been faced with a devastating fire like this one before tonight. We couldn't just quit and go home.

"I'm game if you are," Chuck answered. "We're gonna have to keep each other in sight, though."

This meant that the daisy chain we'd been flying was no longer a good option.

"Roger that," I replied. "You take the lead, and I'll stay on your wing."

Chuck had gained tons of NVG experience in the National Guard, and I trusted him to keep us out of the power lines.

After checking with Kristin to make sure she was okay with the plan to fly formation on Chuck, she and I lowered our goggles from the tops of our helmets and transitioned into the green-shaded world of enhanced night vision. We rendezvoused with Chuck and Scooter as they were climbing out from a drop, and after joining up on their six o'clock, we maintained a loose trail position as we followed them back to the lake.

Once we had refilled them, we lifted our buckets above the dam and began heading toward our first tandem drop on the houses in the cul-de-sac. The fire looked eerily different through the night-vision goggles. The hot spots were clearly visible, but the smoke was just an absence of light. The important thing was that I could clearly see Chuck's aircraft directly in front of mine. As I concentrated on maintaining my position behind him, I had to trust Chuck to see and avoid the power lines directly in our path.

We made more than a dozen passes this way. Initially, I was able to maintain my place in the loose trail formation and still see where I needed to drop, but as we moved further into the night, the winds began to shift and started blowing the smoke back toward us as we made our approaches. In order to keep Chuck in sight, I was forced to move in tighter and tighter on his wing until I was so close that I had to devote all my attention to staying in position.

Unable to look at the target myself, I began using the same technique employed by the bomber crews during World War II. Each time we came across the power lines and dove to make our run on the houses, I kept my eyes glued to Chuck's bird. When he dropped his water—I dropped mine. After we had climbed above the upwind power lines and emerged from the smoke, I would back off into loose trail during the flight back to the lake.

On about the fifth run into the blinding smoke, Kristin came up on the ICS.

"I can't see *anything*," she said. "What are you doing up there?"

"I don't have a clue," I answered flippantly. "You'll have to ask Chuck. He's got the lead."

After flying seven or eight more of these passes in tight formation, I called Chuck on the radio again.

"Are you still feeling good about this?" I asked him.

By this time, the wind had completely turned around. It was rolling the thick smoke down the hillside, making it difficult to see the power lines on our approach to the target—even *with* the NVGs.

"Nope. It's not worth it," Chuck replied.

I was glad to hear I wasn't the only one losing his nerve, and with that, we aborted our attacks on the cul-de-sac and began dropping on the uphill edge of the fire, where Kenny Thompson had been making drops earlier. The smoke wasn't as bad up on that side of the fire now that the wind had shifted, but because we hadn't worked that area in the daylight, we didn't have a good feel for where the obstacles were. As the ground crews vectored us in on some hotspots they wanted us to knock down from the air, Chuck and I made several more drops on that side of the fire.

Finally, as we started to run low on fuel, we had to decide whether to refuel and come back or give up and call it a night.

"I think I'm done," Chuck said over the radio.

"Let's call it," I answered back without hesitating.

The staging area had been moved to a schoolyard, closer to the fire. When we landed, Casey Ping was there with Mike Self. Casey thanked us for the effort and said he concurred with our decision to stand down. In all the years I had known Casey, he had never questioned a pilot's decision not to fly. In spite of our frequent differences, I had always respected him for that.

Chuck Spangler and Scooter White were scheduled to fly the night shift, so they packed up their bird and headed to Brackenridge Hospital. After having fought the fire for the first half of their shift, they ended up flying several EMS missions that night. Oddly enough, all of the "routine" emergencies weren't put on hold just because half the county was going up in flames.

After letting her know she'd done an outstanding job under some extremely demanding conditions, I said goodnight to Kristin McLain—then I hitched a ride home.

There was a small crowd of onlookers when we landed back at the Lakeway Airpark a little before eleven o'clock that night. As he rolled the throttles down to idle, Mark Parcell, who had flown me back from Steiner Ranch, looked at the crowd and asked me if I knew what was up.

"Uh-oh," I said. "I think they're here to complain."

The Lakeway Airpark was a daytime-only airport, and we had gotten complaints in the past for using it as an LZ after dark.

"Good luck with that," Mark said, grinning, as he watched me unstrap and climb down from the copilot seat.

As soon as I was clear, Mark rolled the throttles back up and lifted off, leaving me alone to face the angry neighborhood residents.

To my surprise, they weren't there to complain at all. As soon as the noise from our departing helicopter had subsided, a few in the crowd began clapping, and it wasn't long before the rest of them joined in. The glow from the Steiner Ranch fire was clearly visible from Lakeway, and they had seen television news footage from the fire—footage that showed us making our drops on the burning homes. As I walked toward my car, they converged on me and started telling me how much they appreciated our efforts. I felt a little guilty, especially considering we had given up while the fire was still burning, but I thanked them for their support anyway.

I was too embarrassed to tell them we had abandoned the effort for the night. I also didn't have the courage to tell them we were losing badly in the fight to save the homes in Steiner Ranch.

That was a Sunday night. It wasn't until the following Saturday that the Steiner Ranch fire was finally contained. I spent most of the next two weeks fighting the Pedernales fire. It was the first time since being deployed to the Persian Gulf that I actually got tired of strapping myself into a cockpit.

Day after day, I climbed into the back of an ambulance, along with my fellow STAR Flight crew members, for the ride west on State Highway 71. We were staging our helicopters in a golf course parking lot, located on a hill overlooking the fire. The highway was closed to the public, and as we made our way past the law enforcement checkpoint each day, we girded ourselves for the battle that lay ahead.

Mike Self, who drove the truck that pulled the refueling trailer, was a big part of the fight as well. There was one day, in particular, when we were losing ground to the wildfire, and it had literally surrounded the hill where Mike was stationed. At one point, I was afraid we might have to airlift Mike off the hill and abandon the refueling rig.

Part of the problem was actually finding water to drop on the fire. The drought had left large portions of the Pedernales River completely dry, and we were frequently forced to scoop water from shallow tidal basins in the middle of the riverbed.

Lynn Burttschell was flying in the back of my aircraft that day. In addition to being a good crew chief, Lynn was a volunteer firefighter, and he knew what it took to battle a wildfire. We had to fly nearly three miles from the parking lot to pick up water, and Lynn was getting frustrated with me because our drops were too high. I was trying to stay above the smoke during our runs so I could see what was in front of me, on the other side of the fire line.

Finally, I got tired of Lynn telling me I was too high, so I dove right into the wall of smoke, watching my radar altimeter to maintain my ground clearance. We made our drop just a few feet above the flames, and as we emerged from the smoke on the other side of the fire, I could hear Lynn stomping around in the back of the helicopter.

"What the hell's goin' on back there?" I asked over the ICS.

"Hang on!" he shouted.

A few seconds later, the stomping ceased, and Lynn came up on the ICS again.

"Not *that* low!" he growled indignantly. "We brought the fire with us on that pass."

Some of the burning embers had blown into the back of the aircraft, and Lynn had quickly kicked them back outside as we were climbing out.

Fortunately, we managed to knock down the fire that had surrounded the parking lot where Mike was stationed, but the entire week had been a struggle. We had battled the same fire, day after day, and on every one of those days, we had lost more ground—and homes.

Stemming the Tide

Eventually, about twelve days into the fight, we slowly began to gain the upper hand in the battle to contain the fires. There was nothing we could do at that point to reverse the destruction that had already occurred, of course, so the war itself was still a lost cause. Still—little by little—we began to realize some of the smaller, tactical victories that had eluded us to that point.

During one of those tactical victories, Bill Hanson—the same Bill Hanson who would eventually have doughnuts waiting for me on the night of my concussion—just happened to be flying as my crew chief. In a desperate attempt to save a house, we were dropping directly on top of it, just as Kristin McLain and I had done on that first night of the Steiner Ranch fire. There was also a brush-truck crew on the ground below us. They were laying down water in front of the house as the fire rushed toward it.

Suddenly, the flames hurdled the area they had wetted down, and they were forced to jump in their truck and hightail it out of there. As they were fleeing the fire, their now-unmanned hose was trailing behind the truck, and Bill called them on the radio to let them know they were dragging it on the ground.

They didn't respond, and I told Bill not to worry about it—I was pretty sure they knew they were dragging the hose behind them. The truth of the matter was, they were running for their lives, and securing their loose hose just wasn't very high on their priority list.

Bill and I then watched, helplessly, as the house we were trying to save erupted in flames below us. Unfortunately, it had become an all-too-common theme over the past two weeks.

Bill Hanson and I did have some success that day. There was a house sitting atop a ridge, near the leading edge of the fire. The family was standing outside their home, watching anxiously as the fire climbed the canyon wall toward them. Bill and I tried several times to knock it down, but there was an outcropping above the fire. We were doing a good job wetting down the limestone ledge, but that wasn't doing much to extinguish the fire beneath it.

Finally, I made a high-speed run straight toward the face of the canyon wall and straight toward the fire. From Bill's vantage point, peering out the side of the aircraft, all he could see was the canyon wall looming large in front of us.

"I hope you have a plan here," Bill said with some trepidation in his voice.

"Stand by," I told him.

When we were close enough to the canyon wall, I banked hard to the left and released the water, slinging it toward the fire. I asked Bill if we had hit it, but he said all he could see out of the side door was the sky above us. I shallowed out our turn, and we came around for another look. There was smoke and steam rising from the canyon wall—but no fire. The family was jumping up and down and waving to us, relieved that the flames were no longer advancing on their home.

Hanging out of the aircraft, Bill waved back to them, and then he keyed up the ICS.

"I think they liked that one," he said, chuckling nervously. "Let's not do that anymore."

Bill Hanson and I were one for two that day. In spite of our best efforts, we had lost one house to the fire, but we had successfully defended another.

As we were flying back to the staging area at the end of the day, I detoured over the site of the Steiner Ranch fire to see if the houses Chuck Spangler and I had tried to save were still there. I certainly wasn't going to be surprised if I found them destroyed, and to be honest, by that point, I wasn't sure I would feel any emotion at all if it turned out they had been consumed by the fire. I marked on top of the dam and retraced the same approach we'd flown to the cul-de-sac on that first night of the fire.

When we arrived over the spot where Chuck and I had bombarded the houses, I couldn't believe what I was seeing. Fourteen houses had been reduced to ashes on that street, but right there, in the cul-de-sac below me, there were three—defiantly still standing.

I started laughing out loud, and Bill asked me what I thought was funny.

"Nothing," I said as I quickly sobered. "There's nothing funny about any of this. . . . Let's go home."

Aftermath

The Central Texas wildfires took a huge toll that year. Twenty-three homes were completely destroyed in Steiner Ranch, and a multitude of others were damaged in the fire. Even more tragic, one firefighter lost his life during the effort to contain it. The Pedernales fire scorched nearly seven thousand acres and destroyed more than sixty homes. By far, the largest disaster occurred in Bastrop County, east of Austin. The fires there burned thirty-four thousand acres and destroyed close to seventeen hundred structures. Miraculously, only two people died in that blaze.

A few days after the last of the fires had been extinguished, I was coming off a night shift. I'd been thinking about the massive amounts of water Chuck Spangler and I had dropped on those houses in the cul-de-sac. The water from each one of those drops weighed nearly 2,000 pounds, and I wondered if we'd done any structural damage to the roofs on those homes.

On my way home, I drove out to Steiner Ranch so I could get a good look at the houses from the ground. When I arrived at the entrance to the street that led down the hill to the cul-de-sac, there was a Travis County Sheriff's deputy there, manning a barricade. He stopped me and asked if I lived in the neighborhood. He said only residents were allowed access to the areas with burned-out homes; but then he recognized my STAR Flight uniform, and after I explained what I was doing there, he let me pass.

I parked in the cul-de-sac and got out of my car to take a closer look at the three houses that were left standing. The first thing I noticed was a huge oak tree between the last burned-out home and the first home that had been spared. It was totally charred on the side next to the burned-out home, but it was still green on the side next to the home that had survived the fire.

It's a good thing Chuck was flying lead that night, I thought to myself. *He's a damn' good shot!*

I didn't want to walk into any of the yards, so I stood at the curb, straining to see if there was any visible damage to the roofs on the three houses we had pounded. I guess the woman who lived in the house next to the half-charred oak tree saw me through her window. She must have wondered, *Who is this guy in a blue flight suit staring at my house?*

She came outside and walked toward me.

"Can I help you?" she asked.

I explained who I was, and why I was there. Then I asked her if she had noticed any damage to her roof. She paused for a second, and then she smiled and started laughing.

"That explains a lot," she said. "The insurance guy was here inspecting the house the other day. He said there wasn't any damage, and he also commented that our roof was exceptionally clean, especially considering all the smoke and debris in the neighborhood."

Then she began tearing up slightly and gave me a big hug.

Before I knew it, I had begun tearing up as well—to the point that I was extremely embarrassed about it. I made her promise not to tell anyone she'd seen me crying, at which point she smiled and laughed again, this time through the tears.

"No, seriously," I said. "Please don't let anybody know you saw me react this way. It's bad for my image."

As I was driving past all the burned-out structures on my way out of the neighborhood that day, it eerily reminded me of some familiar historic photographs—the photographs taken of Hiroshima after the blast from the atomic bomb. It occurred to me that it would have taken fifty helicopters to stop the fire that night. In spite of all the feelings of futility I'd experienced over the past couple of weeks, I decided our efforts actually *had* made a difference to somebody.

I called Chuck Spangler to let him know about my conversation with the grateful woman in the cul-de-sac, though I made sure to leave out the part where she and I had both cried like babies. Chuck was pleased, just as I had been, to learn we'd actually done some good on that first night of the fire, and he thanked me for the call. Later that day, I called Kristin McLain and told her as well.

A few weeks later, Kristin and I were flying together again. It was October 12, 2011. The Texas Rangers were playing the Detroit Tigers in game four of the American League Championship Series that night.

It would be my final shift as a STAR Flight pilot.

(above) September, 2011— The Austin skyline is dwarfed by a wall of smoke from the raging wildfires.

(left) The Steiner Ranch evacuation— There was only one road on which to escape the inferno that devastated the community.

A solitary STAR Flight helicopter tries to extinguish a burning home— There were two long weeks during which too many homes were burning, and there simply weren't enough air assets to contain the fires.

Armageddon in Central Texas.

For ten straight days, we rode in the back of an ambulance to reach the staging area on State Highway 71, west of Austin. From there, we flew against the Pedernales fire from dawn to dusk.

(above) An empty bucket swings forward as the helicopter decelerates for a dipping evolution.

(left) Coming up with 2,000 pounds of water—I did this more than three hundred times during the two-week battle, and I was only one of eight STAR Flight pilots who fought against the cataclysmal fires. *Photos Courtesy of Dave Krussow*

Making a pass along the flank—We couldn't halt the fire's advance, so we tried to "herd" it away from structures.

Pilot, Chuck Spangler (left), and crew chief, Scott White (right), were in the lead aircraft during our formation drops on the night of the Steiner Ranch fire.

Chuck Spangler checks out a set of ANVIS 9s, our state-of-the-art night vision goggles.

Lynn Burttschell (left) and Bill Hanson (right) flew as my crew chiefs during the Pedernales fire. Their expertise was invaluable in helping me to cope with the stress and fatigue over those two weeks.

Coming home from the fight.

Epilogue

Once you've tasted flight, you'll forever walk with your eyes turned skyward. There you've been, and there—for as long as you live—you'll long to return.

—Author Unknown

As I close my chronicles, it's been a little more than three years since Kristin McLain and I made the flight that turned out to be my last as a pilot for Travis County. During that time, my fellow STAR Flight aviators have graciously offered to take me flying with them on several occasions, but as of yet, I haven't taken them up on it. I managed to fly for thirty-five years without hurting anybody or bending any airframes, and now that I've retired, I don't feel like pushing my luck any further than I already have. If that makes me appear overly superstitious, so be it. Most pilots *are* superstitious, especially helicopter pilots.

It has certainly been a good life to this point, and I really can't think of anything about it I'd like to change. I dreamed of becoming a naval aviator, and I became one. After that, I dreamed of becoming a STAR Flight pilot, and I got to do that too. Along the way, I married the woman I loved, and we had a daughter and a son, both of whom have made us proud to be their parents. Sure, I guess you could say the concussion that ultimately ended my flying career was a tough break—but just like Forrest Gump, I spent most of my life stumbling into good fortune.

I guess there *is* one thing I would change if I could. If I could go back and do it again, I would tell those closest to me how much I love them more often. My friend Kristin McLain was recently killed during a hoist rescue, and it reminded me just how precious our time on this earth really is. We need to savor every moment we're allowed to spend with family and friends.

I was given a great gift. I was able to parlay my passion for flying into a lifelong career. That storybook career came at a price, however. I witnessed some intense tragedies from the cockpit of my helicopter, scenes that could have shaken me to the core had it not been for my faith in God. I thank Him for giving me the spiritual and emotional stamina that allowed me to move beyond those tragedies and still be able to enjoy flying for the pure, unadulterated pleasure that I derived from it daily.

In stark contrast to the tragedies, I was also witness to some incredible triumphs from my cockpit, and for those experiences, I'm even more grateful. I'll go to my grave knowing that my time here was not merely an existence—it was time well spent.

My life as a pilot has been an incredibly rewarding journey. It's probably no coincidence that some of the most memorable parts of the journey took place *inside the dead man's curve.*

EPILOGUE

Photograph courtesy of Lime Fly Photography, Juan Gonzales.

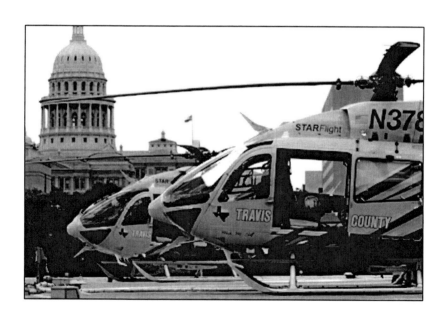

The Letter

On November 15, 2002, exactly one year after Jim Allday, Chris Jones-Piercy, and I pulled Sharon Zambrzycki from the Brushy Creek flood, I came home from a day shift and found the following letter waiting for me on the kitchen table. As I've already recounted, the decision to launch that night was the hardest "go, no-go" decision I ever made. I also stated that I would probably not have made the same decision under similar circumstances later in my career.

That notwithstanding, after reading this letter, I was even more content with my decision to fly that night. I carried the letter as a good-luck charm on every flight I made after receiving it. Two thousand, six hundred and twenty-nine flights later, it was torn and tattered from riding in the left-leg pocket of my flight suit. Fortunately, it was still legible enough for me to transcribe it here:

Dear Kevin, *11-14-02*

Thank you for your courage and determination in rescuing me from Brushy Creek last November. I certainly appreciate your steadfastness. You're a helluva pilot.

This has been a great year, and I treasure each day. Thanks for making it possible.

Congratulations on the Helicopter Heroism Award.

Love, Sharon

Acknowledgments

Many thanks to George Galdorisi for encouraging me and for lending his wonderful expertise to this project. I couldn't have asked for a more accomplished mentor.

Recognition to Scott Spencer of Inkery Studio in Austin, Texas, for expertly illustrating "How the Controls Work" in the front of the book, on page xii.

Special thanks to my editor, Christy Phillippi, and to my sister, Linda Morrow, for tirelessly proofing multiple revisions and helping me to undangle the participles and reunite the infinitives in each and every one of them.

And finally . . .

To all my family, friends, and fellow STAR Flight crew members who painstakingly parsed my manuscripts, checking the many stories for accuracy, thank you for helping me in my effort to "write an honest book."

CPSIA information can be obtained at www.ICGtesting.com
Printed in the USA
LVOW08s1925140816

500371LV00002B/9/P